MATHEMATICAL PHYSICS

MATHEMATICAL PHYSICS

ADVANCED TOPICS

S D Joglekar (Late)
Professor, Department of Physics
Indian Institute of Technology Kanpur
India

Universities Press

MATHEMATICAL PHYSICS: ADVANCED TOPICS

UNIVERSITIES PRESS (INDIA) PRIVATE LIMITED

Registered Office
3-6-747/1/A & 3-6-754/1, Himayatnagar, Hyderabad 500 029, Telangana, India
info@universitiespress.com; www.universitiespress.com

Distributed by
Orient Blackswan Private Limited

Registered Office
3-6-752 Himayatnagar, Hyderabad 500 029, Telangana, India

Other Offices
Bengaluru, Chennai, Guwahati, Hyderabad,
Kolkata, Mumbai, New Delhi, Noida, Patna

First published 2007
Reprinted 2025, 2026

502389

Typeset in Times New Roman *by*
Krishtel eMaging Solutions Pvt. Ltd., Chennai 600 017

ISBN 978 81 7371 560 0

Printed at
Manipal Technologies Limited., Manipal

Published by
Universities Press (India) Private Limited
3-6-747/1/A&3-6-754/1, Himayatnagar,
Hyderabad 500 029, Telangana, India

Contents

Preface

This book is the second part of the two-volume set on mathematical physics written for students at the postgraduate level. Its contents have been determined by my experience of teaching the subject (as well as several other theoretical subjects) to MSc and PhD students at the IIT Kanpur for the last 24years. The two volumes together constitute the material that a student would need in comprehending theoretical courses in physics or chemistry at the postgraduate level. They also constitute a basic prerequisite for a student of physics or chemistry beginning research in theoretical areas.

The first volume entitled *Mathematical Physics: The Basics*, covers the basic knowledge of the field—vectors and tensors, curvilinear coordinates, linear vector spaces and operators, and complex variables and applications. This volume covers more advanced topics such as partial differential equations and ordinary differential equations in physics, special functions including gamma, beta and Bessel functions, classical orthogonal polynomials and spherical harmonics. It also includes the generalized functions, the Dirac-delta function and Fourier transforms. The book closes with a brief introduction to group theory.

Like the earlier voltlme, this one also has several distinctive features. Care has been taken to maintain mathematical rigour throughout, and to encourage intuitive and heuristic understanding of the subject as well. The physical relevance of the topics has been emphasized in each chapter through introductory discussions, examples and illustrations. The book has a large number of solved examples and chapter-end exercises classified according to the main sections. It is tloped that the book's clarity of presentation and the large number of solved examples it includes will help bring the subject within the grasp of every student, and further that the more discerning students and teachers alike would appreciate the book for providing an intuitive understanding of the subject. The book will also be useful to engineering students.

I would like to acknowledge the financial support I received towards the preparation of the book from the IIT Kanpur. It would be an omission on my part if I did not acknowledge the benefits of the meaningful questions posed by some of our students over the years, for these have undoubtedly augmented the clarity of my thinking.

S D Joglekar
IIT Kanpur

25 December 2005

1 Ordinary Differential Equations

1.1 Introduction

As the reader is at least partially familiar, much of Physics involves extensive use of differential equations. A student is first introduced to Physics through Newton's second law which (for a one-dimensional motion) is

$$m\frac{d^2x}{dt^2} = F(x(t), \dot{x}(t), t).$$

This is a second order ordinary differential equation, which when solved for $x(t)$ (with suitable initial conditions) gives the trajectory. Later on, in electrostatics one learns that the electrostatic potential in free space satisfies $\nabla^2\phi = 0$ (the Laplace's equation), and in presence of charge densities, $\nabla^2\phi = -\rho/\epsilon$ (the Poisson equation). The most successful theoretical framework in classical Physics, the Classical Electrodynamics, is formulated entirely in terms of four differential equations, the Maxwell's equations. Out of these as a consequence, follows the wave equation $(\nabla^2 - \frac{1}{c^2}\frac{\partial^2}{\partial t^2})\mathbf{E} = 0$ that implies the existence of electromagnetic waves. Wave equations of similar form also describe waves in elastic media and sound waves. Wave equation, simplified for a sinusoidal time dependence $e^{i\omega t}$, leads to the simpler Helmholtz equation $(\nabla^2 + k^2)\psi = 0$. Entire nonrelativistic Quantum Mechanics is based on the partial differential equation $i\hbar\frac{\partial}{\partial t}\psi = H\psi$, the celebrated Schrödinger equation. Relativistic quantum mechanics also involves the relativistic analogues, viz., the Klein–Gordon and the Dirac equations. All fundamental theories can be expressed in terms of an action S and the consequent Euler–Lagrange equations which are differential equations. Heat transport in gases and materials are expressed in terms of the diffusion equation $\frac{\partial\psi}{\partial t} + k\nabla^2\psi = 0$. Thus, various physical theories are abounding with differential equations in their mathematical formulations.

A little thought would reveal as to why differential equations occur in physics. One of the main motivations behind building a physical theory is to be able to predict the state of a physical system in future, given its state at an initial instant. Thus, one is naturally interested in time derivatives of physical quantities, i.e., how physical quantities change with time. (The reason why time derivatives of physical quantities and not differences in the values of a physical quantity at finite time intervals are generally involved, is related to the continuous manner in which physical quantities are generally thought to change in time.) Thus, the equation of motion of Newton, the Schrödinger equation and some of the Maxwell's equations involve time derivatives of quantities related to a system (position, velocity or wave function or electromagnetic fields, etc.) Now, in an ultimate sense, fundamental theories in physics are relativistically covariant and hence occurrence of space derivatives along with time derivatives is expected. Those theories, then, which are derived from fundamentally relativistic theories as approximations (such as Newtonian mechanics, non-relativistic Schrödinger equation) and the not so fundamental theories based on them are bound to have this feature of having space and time derivatives in their formulations. For this reason, most formulations of physical theories are naturally based on differential equations.

A list of partial differential equations appearing commonly in physics is found in Chapter 5 (see Section 5.1).

1.2 Separation of Variables in Partial Differential Equations

An inspection of the differential equations occurring in the introduction will show that most of them are *partial* differential equations[1]; differential equations which have more than one independent variable (say x, y, z) and partial derivatives with respect to these enter the differential equation. In this chapter we shall mainly be concerned with ordinary differential equations, i.e., the ones in which there is only one independent variable (say x) and ordinary derivatives with respect to it occur in the differential equation. How then do the ordinary differential equations often enter physics? To understand this, we would first like to address a certain method called separation of variables which enables one to simplify a partial differential equation (PDE) into a set of ordinary differential equations (ODE). While this method ideally belongs to the chapter on PDE, logical continuity in understanding the usage of ODE in physics demands that we pursue it here.

An inspection of the PDE's occurring in the introduction will show that most of them contain the partial differential operator ∇^2 the Laplacian. (The Schrödinger equation also generally contains it because when H, the Hamiltonian, is written out in full, the kinetic energy term contains ∇^2). To exhibit the method therefore, we shall choose a simple equation containing the Laplacian ∇^2, viz., the Helmholtz wave equation

$$\left(\frac{\partial^2}{\partial x^2}+\frac{\partial^2}{\partial y^2}+\frac{\partial^2}{\partial z^2}\right)\psi(x,y,z)+k^2\psi(x,y,z)=0, \tag{1.1}$$

and elucidate the method of separation of variables with its help. This can be carried out in many different coordinate systems. While exhibiting the procedure would be simplest in the Cartesian coordinates, we choose instead to carry this out in spherical polar coordinates. This is particularly useful because this is the form in which it is most often needed in

[1] We shall deal with the partial differential equations in greater detail in Chapter 5. There we shall give its precise definition.

physics (and moreover the ODE's that arise out of it are not so trivial unlike with Cartesian coordinates).

Using the result of Chapter 3 of *Mathematical Physics: The Basics** for ∇^2 in spherical polar coordinates, we shall write the equation in terms of r, θ, ϕ as

$$\nabla^2\psi = \frac{1}{r^2\sin\theta}\left[\sin\theta\frac{\partial}{\partial r}\left(r^2\frac{\partial\psi}{\partial r}\right) + \frac{\partial}{\partial\theta}\left(\sin\theta\frac{\partial\psi}{\partial\theta}\right) + \frac{1}{\sin\theta}\frac{\partial^2\psi}{\partial\phi^2}\right]$$
$$= -k^2\psi(r,\theta,\phi). \tag{1.2}$$

To begin with, we make an ansatz: we shall try out a solution ψ which is a product of three functions $A(r)$, $B(\theta)$, $C(\phi)$, each a function of one variable only, viz.,

$$\psi(r,\theta,\phi) = A(r)B(\theta)C(\phi), \tag{1.3}$$

and try to simplify (1.2)[2].

We, now, substitute (1.3) in (1.2) and divide out the equation by ABC to obtain

$$\frac{1}{r^2\sin\theta}\left[\frac{1}{A}\sin\theta\frac{d}{dr}\left(r^2\frac{dA}{dr}\right) + \frac{1}{B}\frac{d}{d\theta}\left(\sin\theta\frac{dB}{d\theta}\right) + \frac{1}{C\sin\theta}\frac{d^2C}{d\phi^2}\right] = -k^2. \tag{1.4}$$

Note that the above equation now contains no partial derivatives but only the ordinary derivatives (of A with respect to r; B with respect to θ and C with respect to ϕ) and thus sets us on the way to obtaining ODE's from the PDE. Multiplying by $r^2\sin^2\theta$ and arranging the terms, we obtain

$$-\frac{1}{C}\frac{d^2C(\phi)}{d\phi^2} = \frac{\sin\theta}{B(\theta)}\frac{d}{d\theta}\left(\sin\theta\frac{dB(\theta)}{d\theta}\right) + \frac{\sin^2\theta}{A(r)}\frac{d}{dr}\left(r^2\frac{dA}{dr}\right) + k^2r^2\sin^2\theta. \tag{1.5}$$

We shall, now, use a very simple argument (that the student will however come upon again and again in his physics studies). Consider the left-hand side of (1.5). It is a function (possibly) only of ϕ. Inspection of the right-hand side, however, will reveal that it is (possible) a function of only r and θ. Now r, θ, ϕ are independent variables. Thus, for example by changing ϕ, right-hand side would not change whereas the left-hand side could possibly change. Thus,

---

[2]Before we proceed, we shall elaborate on the meaning of the ansatz (1.3). We do not imply that every solution of (1.2) is of the form (1.3). In fact, in a physical context, few solutions of (1.2) are likely to be of the form (1.3). One may then think that the method using the above ansatz (1.3) is useful in obtaining only a certain special subset of solution of (1.2). This is not quite true. While solutions of (1.2) of the form (1.3) are indeed a special subset of general solutions of (1.2), it turns out that this subset forms a complete set. Thus, every solution (obeying certain nice mathematical conditions) of (1.2) can be expanded as a linear combination of solutions of the form (1.3). Thus, this ansatz which appears to be very particular in the beginning, leads to the general solution in then end anyhow.

*This is the first part of the two-volume set on mathematical physics by the author.

clearly, (1.5) is meaningful only if the left-hand side is independent of ϕ, i.e., a constant. We shall denote this constant by m^2 for future convenience. (At this stage, no implication is made that m^2 must represent a positive number. However, in applications to physics, it will turn out to be so.) Thus, we have our first ODE

$$\frac{d^2C(\phi)}{d\phi^2} = -m^2C(\phi) \tag{1.6}$$

coming out of (1.2). To continue further, we shall now equate the right-hand side of (1.5) to m^2. This leads us to

$$\frac{\sin^2\theta}{A(r)}\frac{d}{dr}\left[r^2\frac{dA(r)}{dr}\right] + \frac{\sin\theta}{B(\theta)}\frac{d}{d\theta}\left[\sin\theta\frac{dB(\theta)}{d\theta}\right] + k^2r^2\sin^2\theta = m^2. \tag{1.7}$$

Dividing out by $\sin^2\theta$ and rearranging the terms,

$$\frac{1}{A(r)}\frac{d}{dr}\left[r^2\frac{dA(r)}{dr}\right] + k^2r^2 = -\frac{1}{B(\theta)\sin\theta}\frac{d}{d\theta}\left[\sin\theta\frac{dB(\theta)}{d\theta}\right] + \frac{m^2}{\sin^2\theta}. \tag{1.8}$$

We again use the same logic as before. Note that the left-hand side of (1.8) is a function (possibly) of r only, while right-hand side is a function (possibly) of θ only. This is possible only if each side is a constant. We shall denote this constant by $l(l+1)$. (Note that at this stage there is no implication that l is an integer. Every real number can be written in the form $l(l+1)$ for some l.) Having done this we obtain two further ODEs with independent variables θ and r respectively. We have thus shown that by the method of separation of variables, the partial differential equation we started off with are equivalent to three ordinary differential equations listed below.

$$\frac{d}{dr}\left[r^2\frac{dA(r)}{dr}\right] + \left[k^2r^2 - l(l+1)\right]A(r) = 0 \tag{a}$$

$$\frac{1}{\sin\theta}\frac{d}{d\theta}\left[\sin\theta\frac{dB(\theta)}{d\theta}\right] + \left[l(l+1) - \frac{m^2}{\sin^2\theta}\right]B(\theta) = 0 \tag{b}$$

$$\frac{d^2C(\phi)}{d\phi^2} + m^2C(\phi) = 0 \tag{c}$$

The set of solutions of the above three equations (with suitable boundary conditions) will then yield a set of solutions to (1.2) of the form of the ansatz (1.3). By taking suitable linear combinations of these, the general solutions of (1.2) (with suitable boundary conditions) can be constructed.

We shall now briefly recapitulate. Most formulations of physical theories involve differential equations. Usually these are PDEs in space (and time) coordinates. It is generally neither necessary nor easy to attempt to solve the PDE's directly. (For a general treatment of PDEs, see Chapter 5). We can, instead, use the method of separation of variables which seeks solutions that are a product of functions, each a function of only one of the independent variables. This enables us to convert a PDE into a set of ODE's. It is simpler to deal with an ODE as it contains only one independent variable; we shall develop the methods for its solution in this chapter. Given the solutions of this set of ODEs, the set of solutions of PDE of the factorable form [see (1.3)] can be constructed. The entire set of such solutions forms a complete set for solutions of the PDE of physical relevance. Any arbitrary (physical) solution of the PDE with given boundary conditions can then be constructed by taking a linear combination of this set.

Solved problem 1.1 Consider the differential equation

$$\nabla^2\psi(x, y, z) + k^2\psi(x, y, z) = 0$$

satisfied by $\psi(x, y, z)$, a real function.

Apply the method of separation of variables in the Cartesian coordinates. Assume the boundary condition that ψ vanishes on the six faces of a cube of side L.

Solution: We attempt a solution of the form

$$\psi(x, y, z) = X(x)Y(y)Z(z),$$

where X, Y and Z are real functions. The given equation then reads,

$$\left(\frac{\partial^2}{\partial x^2} + \frac{\partial^2}{\partial y^2} + \frac{\partial^2}{\partial z^2}\right)\psi = \frac{d^2X}{dx^2}YZ + \frac{d^2Y}{dy^2}XZ + \frac{d^2Z}{dz^2}XY$$
$$= -k^2XYZ.$$

Now divide out by XYZ to find

$$\frac{1}{X(x)}\frac{d^2X(x)}{dx^2} + \frac{1}{Y(y)}\frac{d^2Y(y)}{dy^2} + \frac{1}{Z(z)}\frac{d^2Z(z)}{dz^2} = -k^2. \quad \text{(i)}$$

Now $\frac{1}{X(x)}\frac{d^2X(x)}{dx^2}$ is a function only of x and the above equation implies that it equals

$$-k^2 - \frac{1}{Y(y)}\frac{d^2Y(y)}{dy^2} - \frac{1}{Z(z)}\frac{d^2Z(z)}{dz^2},$$

which is a function possibly of y and z only. This is possible only when each is a constant. We shall call this constant $-k_1^2$ (this must be a real number, but not necessarily a negative one). Thus,

$$\frac{d^2X}{dx^2} + k_1^2X = 0. \quad \text{(ii)}$$

Further, we have from (i),

$$\frac{1}{Y(y)}\frac{d^2Y(y)}{dy^2} = -k^2 + k_1^2 - \frac{1}{Z(z)}\frac{d^2Z(z)}{dz^2}.$$

Now the left-hand side is a function possibly of y only while the right-hand side is a function possibly of z only. Again, this is possible only if each is a (real) constant which we shall call $-k_2^2$. We thus have

$$\frac{d^2Y(y)}{dy^2} + k_2^2Y(y) = 0,$$

and finally (i) now implies

$$\frac{d^2Z(z)}{dz^2} + (k^2 - k_1^2 - k_2^2)Z(z) \equiv \frac{d^2Z}{dz^2} + k_3^2Z(z) = 0.$$

Now (ii) has the most general solutions

$$X(x) = Ae^{ik_1x} + Be^{-ik_1x}. \tag{iii}$$

Now as k_1^2 is real, k_1 is either real or purely imaginary. We would first like to show that for the present set of boundary conditions, k_1 must be real. To see this, suppose that k_1 is purely imaginary, say $k_1 = iK$, then we would have

$$X(x) = Ae^{-Kx} + Be^{Kx}.$$

Now as $\psi \equiv XYZ$ is given to vanish on the six faces of the cube, $X(x)$ must vanish for $x = 0$ and $x = L$. Thus,

$$A + B = 0 = Ae^{-KL} + Be^{KL}.$$

Thus

$$A(e^{-KL} - e^{KL}) = 0.$$

This implies $A = B = 0$. Thus no solution with given boundary conditions would be possible if k_1 is purely imaginary. A similar conclusion holds also for k_2 and k_3. Thus in (iii), we take k_1 as real. We can then rewrite

$$X(x) = A' \sin k_1x + B' \cos k_1x$$

with A' and B' real (as X is real). Now $X(0) = 0 = X(L)$ implies $B' = 0$ and

$$A' \sin k_1L = 0.$$

Thus, in order to have a nontrivial solution, k_1 can take only discrete values

$$k_1 = \frac{n_1 \pi}{L} \quad n_1 = \pm 1, \pm 2, \ldots$$

(However, in view of the fact that $n_1 = +m$ and $n_1 = -m$ do not yield linearly independent solutions, we can ignore the negative values of k_1.) In a similar manner, we find

$$Y(y) = C \sin k_2 y, \quad k_2 = \frac{n_2 \pi}{L}; \quad n_2 = 1, 2, \ldots$$

$$Z(z) = D \sin k_3 y, \quad k_3 = \frac{\pi n_3}{L}; \quad n_3 = 1, 2, \ldots$$

Therefore the attempted solution

$$\psi(x, y, z) = X(x)Y(y)Z(z)$$

has the form

$$\psi(x, y, z) = \tilde{A} \sin k_1 x \sin k_2 y \sin k_3 z$$
$$k_i = \frac{n_i \pi}{L} \quad n_i = 1, 2, \ldots$$

We note that choosing any of the $n_i's$ equal to zero gives the trivial solution $\psi = 0$. Further, we may restrict all $n_i's$ to be positive since the negative values do not yield new linearly independent solutions.

1.3 Wronskian

In this chapter, we shall deal with the ODE of the following general form

$$y''(z) + p(z)y'(z) + q(z)y(z) = 0 \tag{1.9}$$

which we shall refer to as (I) throughout this chapter. A number of results about this general homogeneous second order ODE with variable coefficients can be derived just from the coefficients $p(z)$ and $q(z)$ in the ODE and without explicitly solving the ODE itself. These results are based on the construction of a quantity called the 'Wronskian' involving two linearly independent solutions of the ODE. With the help of it, we shall be able to show simply the result that there cannot be more than two linearly independent solutions to such an ODE.

We shall also be able to obtain the explicit value for the Wronskian (up to an overall constant) in terms of p and q only[3] *without knowing the explicit form* of the linearly independent solutions. This object, thus immediately evaluable from the ODE, has many applications, the major one being that it enables one to obtain a second linearly independent solution if one is known already by some method or by inspection.

[3]Actually, the Wronskian does not involve $q(z)$ at all.

1.3.1 Linear dependence of functions

The concept of linear independence of functions is quite analogous to that of linear independence of vectors in a linear vector space (and in fact a special case of it. See below).

Consider a set of n functions: $\{y_1(z), \dots y_n(z)\}$, each defined for every $z \in D$, D being a domain in a complex plane. The n functions are said to be linearly dependent[4] if there exist constants $a_1, a_2, \dots, a_n$, not all zero, such that

$$\sum_{i=1}^{n} a_i y_i(z) = 0; \quad \forall z \in D.$$

While the definition of linear dependence does not *require* it, in the following we shall assume that $y_i(z)$, $i = 1, 2, \dots, n$, are differentiable in D as often as we would like it in a given context. It turns out to be sufficient for our purpose.

A necessary condition for linear dependence of n functions

Consider a set of n functions $y_1(z), \dots, y_n(z)$ defined for $z \in D$. Consider a point $z_0 \in D$ at which all derivatives up to order $(n-1)$ are well defined for all y_i's. Now, if the n functions are linearly dependent, then there exists n constants $a_1, \dots a_n$, not all zero, such that

$$\sum_{i=1}^{n} a_i y_i(z) \equiv 0; \quad \forall z \in D. \tag{1.10}$$

We can differentiate this relation s times $(1 \le s \le n-1)$. And set $z = z_0$. Denoting by $y_i^{(s)}(z)$ the sth derivative of $y_i(z)$ (with the convention that $y_i^{(0)}(z) = y_i(z)$), we arrive at n homogeneous equations.

$$\sum a_i y_i^{(s)}(z_0) = 0; \quad s = 0, 1, \dots, (n-1) \tag{1.11}$$

in n 'variables' $a_1, \dots a_n$. These can be written in the matrix form

$$\begin{bmatrix} y_1(z_0) & y_2(z_0) & \dots & y_n(z_0) \\ y_1^{(1)}(z_0) & y_2^{(1)}(z_0) & & y_n^{(1)}(z_0) \\ \vdots & \vdots & & \vdots \\ y_1^{(n-1)}(z_0) & y_2^{(n-1)}(z_0) & & y_n^{(n-1)}(z_0) \end{bmatrix} \begin{bmatrix} a_1 \\ a_2 \\ \vdots \\ a_n \end{bmatrix} = 0. \tag{1.12}$$

Since the column vectors on the left-hand side consisting of a_i's is not identically a null vector, this requires that the determinant of the $n \times n$ matrix on the left-hand side is identically zero. That is,

[4]We can look upon the following definition of linear dependence as a special case of that of a linear vector space. This linear vector space (infinite-dimensional) is of functions of a complex variable defined over a domain D. A function $f(z)$ defined by $f(z) = 0 \, \forall z \in D$ is the null vector in this linear vector space. The above is just the definition of linear dependence of n vectors $\{y_i(z) : i = 1, 2, \dots, n\}$ in a vector space.

$$W(z_0) \equiv \begin{vmatrix} y_1(z_0) & y_2(z_0) & \cdots & y_n(z_0) \\ y_1^{(1)}(z_0) & y_2^{(1)}(z_0) & & y_n^{(1)}(z_0) \\ \vdots & \vdots & & \vdots \\ y_1^{(n-1)}(z_0) & y_2^{(n-1)}(z_0) & & y_n^{(n-1)}(z_0) \end{vmatrix} = 0 \tag{1.13}$$

This determinant is called the 'Wronskian' of n functions, $y_1, y_2, \ldots, y_n$. The Wronskian of n linearly dependent functions vanishes at every $z \in D$.

Conversely, consider a set of n functions $v_1(z), \ldots v_n(z)$, defined over D. To test for linear independence we would write a relation

$$\sum_{i=1}^{n} b_i v_i(z) = 0 \quad \text{for every} \quad z \in D \tag{1.14}$$

and see if this necessarily implies all $b_i = 0$.

Suppose, now, that v_i's are all $(n-1)$ times differentiable at a point $z = z_0 \in D$, one would end up with an equation of the type (1.12) for coefficients b_i's.

Now, if the Wronskian of the n functions W is nonzero at this point z_0, then this necessarily implies that all $b_i = 0$. This then implies that the n functions $v_1, \ldots, v_n$ are linearly independent over D.

Thus if the Wronskian of n functions is defined over D and is *nonzero* at *some point* in D, then the n functions are linearly independent in D.

Note that the above is a *sufficient* (but not necessary) condition for linear independence. (See, for example, solved Problem 1.4 in Section 1.3.2.)

1.3.2 Number of linearly independent solutions to a second order linear homogeneous ODE

Consider the linear homogeneous ODE of second order with variable coefficients. We shall assume that the singularities of the coefficient functions in the ODE are a finite number of poles, so that by multiplying by a suitable factor it can be recast in the form

$$R(z)\frac{d^2y}{dz^2} + P(z)\frac{dy}{dz} + Q(z)y = 0; \quad \forall z \in D \tag{1.15}$$

with $P(z)$, $Q(z)$ and $R(z)$ analytic at each $z \in D$.

We shall also assume, without loss of generality, that at every point $z \in D$ that all of R, P, Q do not simultaneously vanish. (Why?)

Now let $y_1(z)$, $y_2(z)$, and $y_3(z)$, be three solutions to the ODE, assumed to exist. We want to show that they cannot be linearly independent. We express the fact that y_1, y_2 and y_3 are solutions of (1.15) at $z = z_0$ by the following three equations.

$$R(z_0)y_1^{(2)}(z_0) + P(z_0)y_1^{(1)}(z_0) + Q(z_0)y_1(z_0) = 0$$
$$R(z_0)y_2^{(2)}(z_0) + P(z_0)y_2^{(1)}(z_0) + Q(z_0)y_2(z_0) = 0$$
$$R(z_0)y_3^{(2)}(z_0) + P(z_0)y_3^{(1)}(z_0) + Q(z_0)y_3(z_0) = 0 \tag{1.16}$$

These equations together can be written in a matrix form as

$$\begin{bmatrix} y_1^{(2)}(z_0) & y_1^{(1)}(z_0) & y_1(z_0) \\ y_2^{(2)}(z_0) & y_2^{(1)}(z_0) & y_2(z_0) \\ y_3^{(2)}(z_0) & y_3^{(1)}(z_0) & y_3(z_0) \end{bmatrix} \begin{bmatrix} R(z_0) \\ P(z_0) \\ Q(z_0) \end{bmatrix} = 0. \tag{1.17}$$

Now, by our assumption, the elements of the column vector on the left-hand side do not all vanish. Hence the determinant of the 3×3 matrix, which is the Wronskian of y_1, y_2 and y_3, must vanish at z_0. This is valid for each point $z_o \in D$. This implies that the three functions y_1, y_2, y_3 are *linearly dependent* over D.

This proves that there are at most two linearly independent solutions to an ODE (1.15). Later on, we shall deal with the results regarding the existence of solutions to the ODE (I) in the vicinity of a point $z = z_0$, with some restrictions on the possible singularities of p and q at z_0. We shall then establish the existence of two linearly independent solutions to (I).

Solved problem 1.2 Verify directly that the three functions $f_n(z) = z^n$ $n = 0, 1, 2$ are linearly independent over the complex plane $\mathbb{C}$. Evaluate the Wronskian and show that it implies the linear independence of the three functions.

Solution: The three functions f_0, f_1, f_2 are linearly independent over the complex plane if a relation of the form

$$a_0 f_0 + a_1 f_1 + a_2 f_2 = 0 \tag{i}$$

over $\mathbb{C}$ implies necessarily implies $a_0 = a_1 = a_2 = 0$. Now, (i) reads

$$a_0 + za_1 + z^2 a_2 = 0. \tag{ii}$$

We now set $z = 0$ in (ii). We have

$$a_0 = 0. \tag{iii}$$

Then, (ii) implies

$$za_1 + z^2 a_2 = 0 \Rightarrow a_1 + za_2 = 0, \quad z \neq 0. \tag{iv}$$

Now, set $z = \pm 2$ in (iv). We then have

$$a_1 + 2a_2 = 0$$

and

$$a_1 - 2a_2 = 0. \tag{v}$$

These further leads to $a_1 = a_2 = 0$. This proves directly the linear independence of the three functions over D.

The Wronskian W is

$$W = \begin{bmatrix} f_0 & f_1 & f_2 \\ f_0' & f_1' & f_2' \\ f_0'' & f_1'' & f_2'' \end{bmatrix} = \begin{bmatrix} 1 & z & z^2 \\ 0 & 1 & 2z \\ 0 & 0 & 2 \end{bmatrix} = 2. \tag{1.18}$$

$W \neq 0$ everywhere in D.

This necessarily implies that the three functions are linearly independent over D.

Solved problem 1.3 Verify the linear independence of the first three Hermite polynomials. $H_0 = 1$, $H_1 = 2x$, $H_2 = 4x^2 - 2$.

Solution:

$$W = \begin{vmatrix} H_0 & H_1 & H_2 \\ H_0' & H_1' & H_2' \\ H_0'' & H_1'' & H_2'' \end{vmatrix} = \begin{vmatrix} 1 & 2z & 4z^2 - 2 \\ 0 & 2 & 8z \\ 0 & 0 & 8 \end{vmatrix} = 16 \tag{1.19}$$

$W \neq 0$ everywhere in $\mathbb{C}$. Hence H_0, H_1 and H_2 are linearly independent over $\mathbb{C}$.

Note that we had already demonstrated in Problem 1.2 that $1, z, z^2$, are linearly independent over C. The three Hermite polynomials are linearly independent combinations of these, since

$$\begin{bmatrix} H_0 \\ H_1 \\ H_2 \end{bmatrix} = \begin{bmatrix} 1 & 0 & 0 \\ 0 & 2 & 0 \\ -2 & 0 & 4 \end{bmatrix} \begin{bmatrix} 1 \\ z \\ z^2 \end{bmatrix} \quad \text{with} \quad \begin{vmatrix} 1 & 0 & 0 \\ 0 & 2 & 0 \\ -2 & 0 & 4 \end{vmatrix} = 8 \neq 0 \tag{1.20}$$

and hence themselves linearly independent.

Solved problem 1.4 Show that the functions $f_1(z) = \operatorname{Re} z$ and $f_2(z) = \operatorname{Im} z$ are linearly independent over D. Show nonetheless that the Wronskian is not defined anywhere in D.

Consider the relation

$$af_1 + bf_2 = 0.$$

i.e., $ax + by = 0 \qquad z \in D$.

Consider the above relation at the points (0,1) and (1,0). We thus obtain

$$a = b = 0.$$

proving linear independence.

f_1 and f_2 are not analytic anywhere in D, hence the Wronskian does not exist anywhere.

(This example brings out the fact that the concept of linear independence of n functions is defined irrespective of the existence of the Wronskian. The Wronskian test is a *sufficient* condition for testing linear independence but not a necessary one.)

1.3.3 Properties of the Wronskian of two functions

We shall note a few simple properties of the Wronskian of two functions y_1 and y_2 that follow from the definition.

(a) $W[y_1, cy_1] \equiv 0$

(b) $W[y_1, y_2 + cy_1] = W[y_1, y_2] + W[y_1, cy_1] = W[y_1, y_2]$

(c) $W[\alpha_1 y_1 + \alpha_2 y_2, \beta_1 y_1 + \beta_2 y_2] = (\alpha_1\beta_2 - \alpha_2\beta_1)W[y_1, y_2]$

(d) $W[y_1, y_2] = 0$ and $y_1 \neq 0$ over a domain D necessarily implies $y_2 = cy_1$ over D.

The proofs of these simple results are left as exercises (see Exercise 1.2).

Solved problem 1.5

(i) Find the Wronskian of $y_1 = z^2$ and $y_2 = \sin z$.

(ii) Find the Wronskian of $\tilde{y}_1 = z^2 + \sin z$ and $\tilde{y}_2 = z^2 - 5\sin z$.

(iii) Verify that the two Wronskians differ by a constant factor. Show that this constant factor is compatible with that in (c) above.

Solution:

(i)

$$W_1 = \begin{vmatrix} y_1 & y_2 \\ y_1' & y_2' \end{vmatrix} = \begin{vmatrix} z^2 & \sin z \\ 2z & \cos z \end{vmatrix}$$

$$= z^2 \cos z - 2z \sin z$$

(ii)

$$W_2 = \begin{vmatrix} \tilde{y}_1 & \tilde{y}_2 \\ \tilde{y}_1' & \tilde{y}_2' \end{vmatrix} = \begin{vmatrix} z^2 + \sin z & z^2 - 5\sin z \\ 2z + \cos z & 2z - 5\cos z \end{vmatrix}$$

$$= (z^2 + \sin z)(2z - 5\cos z) - (z^2 - 5\sin z)(2z - \cos z)$$

$$= -6W_1$$

We note that $\tilde{y}_1 = \alpha_1 y_1 + \alpha_2 y_2$ and $\tilde{y}_2 = \beta_1 y_1 + \beta_2 y_2$ with $\alpha_1 = \alpha_2 = 1 = \beta_1$ and $\beta_2 = -5$. Hence the factor $(\alpha_1\beta_2 - \beta_1\alpha_2) = -6$ as expected from the property (c) above.

1.3.4 The Wronskian method for finding a second linearly independent solution

In the next two sections (1.5–1.6), we shall develop some general results on the existence of and the number of linearly independent solutions of ODE under consideration. There, we will show the existence of two linearly independent solutions of the general ODE (I)

$$y'' + py' + qy = 0,$$

and develop techniques for actually constructing two linearly independent solutions. (This will be done subject to certain restrictions on p and q). In many cases, however, a short-cut to the construction of a solution is available. Sometimes it is possible to read out, by inspection, one of the solutions of (I). Alternately, one may have a technique available for developing one of the solutions. In such cases the Wronskian method enables one to construct a *second* linearly independent solution. This method, moreover, will also help us in the development of general results regarding the nature of solutions in the vicinity of what is called 'a regular singular point'. This will be dealt with in detail in the Section 1.6.

Examples

(i) Consider the ODE

$$y'' + zy' - y = 0.$$

An inspection of the same allows us to immediately write one solution: $y = z$.

(ii) Consider the ODE

$$z^2y'' + zy' - 4y = 0.$$

The above equation is homogeneous in z: a change of independent variable $z \to \alpha z$ (α is a nonzero constant) leads us to the same equation. We therefore expect a solution that scales (at most) by a constant under this change. This suggests that we try $y = z^m$. Then, we have

$$m(m-1) + m - 4 = 0 \Rightarrow m = \pm 2.$$

This method, in fact, gives both the linearly independent solutions: $y = z^{\pm 2}$.

(iii) For the Hermite differential equation

$$y'' - 2zy' + 2\alpha y = 0,$$

one solution, for $\alpha = 0$, is easily seen to be $y_1 = 1$.

The basic result behind this method is that given *any* two solutions y_1 and y_2 of the ODE, the form of the Wronskian is identical (up to a constant), which is further determined only by the function $p(z)$ appearing in (I). In other words,

$$W[y_1, y_2] = W[z] = Cf(z)$$

and

(i) taking different pairs (y_1, y_2) of solutions of the ODE simply alters the constant C and not the functional form $f(z)$;
(ii) $f(z)$ depends *only* on $p(z)$.

To see the property (i), let us suppose that (y_1, y_2) are a pair of linearly independent solutions of (I). Consider any other pair $\tilde{y}_1$ and $\tilde{y}_2$ of the solutions of (I). As there are at the most two linearly independent solutions to (I), there must exist $\alpha_1, \alpha_2, \beta_1$ and β_2 such that

$$\tilde{y}_1 = \alpha_1 y_2 + \alpha_2 y_2,$$
$$\tilde{y}_2 = \beta_1 y_2 + \beta_2 y_2.$$

Thus, from the property (c) of W listed in Section 1.3.3,

$$\begin{aligned} W[\tilde{y}_1, \tilde{y}_2] &= (\alpha_1\beta_2 - \alpha_2\beta_1)W[y_1, y_2] \\ &= (\alpha_1\beta_2 - \alpha_2\beta_1)Cf(z) \\ &= C'f(z). \end{aligned}$$

This shows that the Wronskian of any two solutions of the ODE has the same functional dependence.

We shall now explicitly construct the function $f(z)$ and show how it depends only on $p(z)$, which will confirm property (ii) above. Let y_1 and y_2 be any two solutions of the ODE (I) in domain D. Then[5],

$$y_1'' + p(z)y_1' + q(z)y_1 = 0, \qquad \text{(i)}$$
$$y_2'' + p(z)y_2' + q(z)y_2 = 0. \qquad \text{(ii)}$$

We multiply the equation (i) by y_2 and (ii) by y_1 and subtract. This gives

$$(-y_1''y_2 + y_2''y_1) + p(z)(-y_1'y_2 + y_2'y_1) = 0. \qquad \text{(iii)}$$

We note that the q-dependent terms are no longer present in (iii). Now, we recognise the second term on the left-hand side as $W(y_1, y_2)$, and the first term as

$$y_1y_2'' - y_1''y_2 = \frac{d}{dz}(y_1y_2' - y_2y_1') = \frac{dW}{dz}.$$

[5]In writing down the following equations for any $z \in D$, we are in effect assuming the existence of y_1' and y_2' throughout D. In other words, we are *a priori* assuming analytic nature of the solutions in D. (It also amounts to an assumption about $p(z)$: Eq. (iii) below implies that $p(z)$ must be a ratio of two analytic functions). In the future applications of this result, we will be able to justify the assumptions. Alternately, one can as well derive the following result with a *real* independent variable x.

Thus (iii) can be rewritten as

$$\frac{dW}{dz} + p(z)W(z) = 0.$$

This equation can be easily integrated to yield

$$W[z] = W[z_0]\exp\left\{-\int_{z_0}^{z} p(z')dz'\right\}, \tag{1.21}$$

where the path from z_0 to z lies in D. W is analytic over the domain has been assumed.

Solved problem 1.6 Find the Wronskian (up to an arbitrary constant) for the two linearly independent solutions of the Bessel equation

$$y'' + \frac{1}{z}y' + \left(1 - \frac{\nu^2}{z^2}\right)y = 0.$$

Does the Wronskian depend upon ν?

Solution: Here[6],

$$p(z) = \frac{1}{z}; \quad q(z) = 1 - \frac{\nu^2}{z^2}; \quad D = \mathbb{C} - \{0\}$$

$$\begin{aligned} W[z] &= W[z_0]\exp\left\{-\int_{z_0}^{z} p(z')dz'\right\} \\ &= W[z_0]\exp\left\{-\ln\frac{z}{z_0}\right\} = W[z_0]\frac{z_0}{z} \equiv \frac{C}{z} \quad z \neq 0. \end{aligned}$$

Thus, we find that the Wronskian of two linearly independent solutions of the Bessel equation must be of the form $\frac{C}{z}$; (with $C \neq 0$ on account of linear independence).

We add a number of observations:

1. The result for W does not depend on ν. Dependence on ν could only have arisen through $q(z)$ and W is assured to be independent of $q(z)$.
2. We also note that while $\ln\left(\frac{z}{z_0}\right)$ is a multivalued function, the result for W involves $\exp\left\{-\ln\left(\frac{z}{z_0}\right)\right\}$ which is a single-valued function as $\exp(-2\pi i n) = 1$.
3. We stress that, *even before one has attempted to solve* the Bessel equation and *even before one has any knowledge of the form* of its two linearly independent solutions, one already knows the form of

$$W[y_1, y_2] = y_1 y_2' - y_2 y_1' = \frac{C}{z}.$$

[6] The statement below regarding the domain D implies that D is the complex plane with origin $z = 0$ omitted.

For this, we only need to know $p(z) = \frac{1}{z}$. Evidently, this is a simple yet powerful result.

To appreciate this further, we illustrate this result with the following example.

Consider again the Bessel differential equation

$$y'' + \frac{1}{z}y' + \left(1 - \frac{\nu^2}{z^2}\right)y = 0 \tag{i}$$

and the following ODE

$$y'' + \frac{1}{z}y' - 4y = 0. \tag{ii}$$

We note that solutions of (i) are *specific infinite* series (we shall show this in Section 1.6) in z called the Bessel functions (see Chapter 3). We have already seen that the solutions to (ii) are simple, viz., $y_1 = z^2$ and $y_2 = z^{-2}$. Thus the solutions to the differential equations (i) and (ii) have no resemblance to each other whatever. Nonetheless, the specific construct depending on two linearly independent solutions, viz., the Wronskian, however turn out to have the same from $\frac{C}{z}$ in both cases. The *detailed form* of y_1 and y_2 very much depends on $q(z)$, which is different in (i) and (ii). But W depends only on $p(z)$ which is same in (i) and (ii).

Determination of a second linearly independent solution by the Wronskian method

Suppose one were to know one solution of the ODE at hand by some means (say by inspection or by the series solution method, to be discussed later), then this information together with the knowledge of the Wronskian from (1.21) gives us

$$y_1y_2' - y_1'y_2 = W[z] = \text{ a quantity } \textit{known} \text{ up to a constant.}$$

The above is a *first* order differential equation for y_2 and can be solved by the method of integrating factor. We note

$$\frac{y_1y_2' - y_1'y_2}{y_1^2} = \frac{d}{dz}\left(\frac{y_2}{y_1}\right) = \frac{W[z]}{y_1^2(z)}$$

provided y_1 is nonzero at a point z.

Integrating the above result from z_0 to z along a path in D on which $y_1 \neq 0$ anywhere, we have

$$y_2(z) = y_1(z)\int_{z_0}^{z} \frac{dz'}{y_1(z')^2} W(z') \tag{1.22}$$

$$= y_1(z)\int_{z_0}^{z} \frac{dz' W(z_0)}{y_1(z')^2} \exp\left[-\int_{z_0}^{z'} p(z'')dz''\right]. \tag{1.23}$$

We now make sure that we have chosen the point z_0 such that $W[z_0] \neq 0$. Thus, $y_2(z)$ above gives a second linearly independent solution of the ODE in domain D.

We can simplify the above expression for a second linearly independent solution by noting the following:

1. If y_2 is linearly independent of y_1 so is Dy_2, where D is a nonzero constant.
2. If y_2 is linearly independent of y_1, than so is $y_2 + Ey_1$, where E is a constant.

We can, thus, use (1) to remove the constant factor $W[z_0]$ from (1.23). Further, we note that changing the lower limit $z_0 \to \tilde{z}_0$ in the second integral is compensated by the the change $W(z_0) \to W(\tilde{z}_0)$ (already dealt with) while a similar change in the first integral only changes y_2 by

$$y_1 \int_{\tilde{z}_0}^{z_0} \frac{dz'}{y_1(z')^2} \exp\left[-\int_{\tilde{z}_0}^{z'} p(z'')dz''\right] \equiv (\text{const})y_1.$$

Thus, using (2) above, we can, in effect, ignore the contribution from the lower end. Thus, we can choose the expression for the second linearly independent solution as

$$y_2(z) = y_1(z) \int^{z} \frac{dz'}{y_1(z')^2} \exp\left[-\int^{z'} p(z'')dz''\right], \tag{1.24}$$

where we regard the two integrals on the right-hand side as *indefinite* integrals.

Solved problem 1.7 Find a second linearly independent solution of the ODE

$$y'' + 2zy' - 2y = 0$$

given that one of the solutions is $y_1 = z$.

Solution: By inspection, we verify that $y_1 = z$ is a solution of the given ODE. A linearly independent second solution $y_2(z)$ can thus be taken as

$$y_2(z) = y_1(z) \int^{z} \frac{dz'}{y_1^2(z')} \exp\left[-\int^{z'} p(z'')dz''\right].$$

Here $p(z') = 2z'$,

$$y_2 = y_1 \int^{z} \frac{dz'}{z'^2} \exp(-z'^2)$$

$$= y_1(z) \int^{z} \frac{dz'}{z'^2}\left[1 - z'^2 + \frac{1}{2}z'^4 - \ldots\right]$$

$$= z\left[-\frac{1}{z} - z + \frac{z^3}{6} - \cdots\right]$$

$$= -1 - z^2 + \frac{z^4}{6} + \cdots$$

We note that in the expression for $y_2(z)$ we are ignoring the lower limit in each of the two integrals. This is consistent with what was elaborated previously.

Solved problem 1.8 Consider the ODE $y'' + \frac{1}{z}y' - \frac{m^2}{z^2}y = 0$.

(a) Show by inspection that $y_1 = z^m$ is one solution of the above equation.

(b) Apply the Wronskian method to obtain a second linearly independent solution.

Solution:

$$y_1' = mz^{m-1}$$

$$y_1'' = m(m-1)z^{m-2}$$

$$y_1'' + \frac{1}{z}y_1' - \frac{m^2}{z^2}y_1 = \left[m(m-1) + m - m^2\right]y = 0$$

Thus, $y_1 = z^m$ is one solution of the ODE. Now, the Wronskian method gives a second linearly independent solution as

$$y_2 = y_1 \int^z \frac{dz'}{y_1(z')^2} \exp\left[-\int^{z'} p(z'')dz''\right]$$

$$= y_1(z) \int^z \frac{dz'}{z'^{2m}} \exp\left[-\ln z'\right]$$

$$= y_1(z) \int^z \frac{dz'}{z'^{(2m+1)}}$$

$$= y_1(z)\frac{1}{z^{2m}}\left(\frac{-1}{2m}\right) \qquad m \neq 0$$

$$= z^{-m} \times (\text{const}).$$

Thus, we can take $y_2 = z^{-m}$ as the second solutions provided $m \neq 0$. On the other hand, if $m = 0$, we have, $y_1 = 1$ and

$$y_2 = y_1 \int^z \frac{dz'}{z'} = \ln z.$$

1.4 Nature of Solutions of the ODE: General Discussion

We now consider the ODE in the most general form

$$\frac{d^2y(z)}{dz^2} + p(z)\frac{dy(z)}{dz} + q(z)y(z) = 0, \tag{1.25}$$

where the functions $p(z)$ and $q(z)$ are specified in a domain D in the complex plane. We want to solve (I) about a point $z = z_0 \in D$; i.e., we want the general solution to (I) in the vicinity of $z = z_0$. It stands to reason (and it will soon be discussed extensively) that the analyticity properties of the general solution $y(z)$ at $z = z_0$ depend on the analyticity properties of $p(z)$ and $q(z)$ at $z = z_0$. We want to classify now what the precise consequences of different analytic properties (i.e., whether p and q are analytic, have a pole of nth order or an essential singularity etc.) of $p(z)$ and $q(z)$ at $z = z_0$ are. With a view to formulate these consequences and also for the sake of future developments of the subject matter, we wish to cast (1.25) in a simpler form by a change of the dependent variable.

We consider

$$y(z) \equiv \exp\left[-\int_a^z \frac{1}{2}p(z')dz'\right] f(z) \equiv I(z)f(z) \tag{1.26}$$

where the integral in the exponent is chosen along any path from an arbitrary point a to z through the region of analyticity of $p(z)$. Then,

$$y'(z) = -\frac{1}{2}p(z)y(z) + I(z)f'(z),$$

$$y''(z) = -\frac{1}{2}p(z)y'(z) - \frac{1}{2}p'(z)y(z) - \frac{1}{2}p(z)I(z)f'(z) + I(z)f''(z)$$

$$= -p(z)y'(z) - \frac{1}{2}p'(z)y(z) - \frac{1}{4}p^2(z)I(z)f(z) + I(z)f''(z),$$

so that (I) becomes (recall that $I(z) \neq 0$)

$$\frac{d^2f(z)}{dz^2} + Q(z)f(z) = 0 \tag{1.27}$$

with

$$Q(z) = q(z) - \frac{1}{4}p^2(z) - \frac{1}{2}p'(z). \tag{1.28}$$

[We shall henceforth refer to (1.27) as (II).] We shall now employ (II) to carry out a study and to classify the analytic properties of the general solution to (I). From (1.26) we note that the analytic properties of $y(z)$ will generally depend on those of both the functions $I(z)$ and $f(z)$.

Further, as seen from the definition of $I(z)$ of (1.26), the analytic properties of $I(z)$ can depend on only those of $p(z)$. Also, from (1.27) we see that the analytic properties of a solution $f(z)$ of it can depend on $Q(z)$.

In this section, we wish to establish two simple results about the precise manner in which the analytic properties of $I(z)$ depend upon those of $p(z)$ and the necessary conditions on $Q(z)$ for an analytic solution $f(z)$ of (1.27) to exist.

Theorem 1.1 Let $p(z)$ have a pole of order m at $z = z_0$ with a residue b_1. Then,

(i) for $m = 1$, $I(z)$ has a branch point at $z = z_0$ unless $-\frac{b_1}{2}$ = a non-negative integer,

(ii) for $m > 1$, $I(z)$ has a factor having an essential singularity at $z = z_0$, and a factor having a branch-point there unless $-\frac{b_1}{2}$ = is a non-negative integer.

Proof: Since $p(z)$ has a pole of order m at $z = z_0$, a Laurent expansion of the form

$$p(z) = \frac{b_m}{(z - z_0)^m} + \cdots + \frac{b_1}{(z - z_0)} + p_0(z) \qquad b_m \neq 0$$

exists in an annulus $0 < |z - z_0| < R(R > 0)$ with $p_0(z)$ analytic in the disc $|z - z_0| < R$. We take z, an arbitrary point a and the path joining them all in the annulus $0 < |z - z_0| < R$; we have

$$\int_a^z p(z')dz' = \int_a^z \left[\sum_{r=1}^{m} \frac{b_r}{(z' - z_0)^r} + p_0(z')\right] dz'$$

$$= \sum_{r=2}^{m} \frac{-b_r}{(r-1)(z - z_0)^{r-1}} + b_1 \ln(z - z_0) + g(z)$$

where[7] $g(z)$ is analytic in $|z - z_0| < R$. Thus,

$$I(z) = \exp\left\{\frac{1}{2}\sum_{r=2}^{m} \frac{b_r}{(r-1)(z - z_0)^{r-1}} - b_1\frac{1}{2}\ln(z - z_0) - \frac{1}{2}g(z)\right\}$$

$$= \exp\left\{\sum_{r=2}^{m} \frac{b_r}{2(r-1)(z - z_0)^{r-1}}\right\}(z - z_0)^{-b_1/2}\exp\left\{-\frac{1}{2}g(z)\right\}.$$

Now we note the analytic properties of each of the factors:

1. $\exp\left\{\sum_{r=2}^{m} \frac{b_r}{2(r-1)(z-z_0)^{r-1}}\right\}$ has an essential singularity at $z = z_0$ for $m \geq 2$
2. $(z - z_0)^{-b_1/2}$ has a branch cut at $z = z_0$ unless $-\frac{b_1}{2}$ is a non-negative integer.
3. $\exp\left(-\frac{1}{2}g(z)\right)$ is analytic at $z = z_0$.

An inspection of these observations enables one to draw the conclusions in the theorem.

[7] We have absorbed appropriate constant terms into $g(z)$.

Examples Consider (i) $p(z) = \frac{1}{z}$, (ii) $p(z) = \frac{1}{z^2}$, (iii) $p(z) = 2$. In these three cases, we have the result for $I(z)$ as

(i) $\sqrt{\frac{a}{z}}$

(ii) $\exp\{\frac{1}{2z} - \frac{1}{2a}\}$

(iii) $\exp(-z+a)$ and these respectively have (i) has a branch-cut at $z = 0$, (ii) an essential singularity at $z = 0$, (iii) a point of analyticity at $z = 0$.

Theorem 1.2 Eq. (1.27) can have a solution that is analytic at $z = z_0$ only if Q has no worse (singularity) than a double pole, i.e., a Laurent series for Q of the form

$$Q(z) = \frac{b_2}{(z - z_0)^2} + \frac{b_1}{z - z_0} + q(z)$$

exists in some annulus $0 < |z - z_0| < R$, with $q(z)$ analytic in the disk $|z - z_0| < R$.

Proof: Suppose a solution $f(z)$ of (1.27) analytic at $z = z_0$ exists. Then we can write the Taylor expansion for $f(z)$ convergent in some disc $|z - z_0| < R_1$ as

$$f(z) = \sum_{n=0}^{\infty} a_n (z - z_0)^n.$$

From (1.35), we can express $Q(z)$ in the domain of validity of the solution $f(z)$ as

$$Q(z) = -f''(z)/f(z) \qquad \text{(i)}$$

provided $f(z) \neq 0$. Now we consider the two cases separately.

Case (i): $f(z_0) \neq 0$. Then, as $f(z)$ is analytic in $|z - z_0| < R_1$, the nearest zero of $f(z)$, if any, is a finite distance (R_2) away from z_0. Let $R_3 = \min(R_1, R_2)$. Then in the disc $|z - z_0| < R_3$, $f''(z)$ is analytic and hence $Q(z)$ given by (i) is also analytic.

Case (ii): Let $f(z)$ have a zero of order m at $z = z_0$. We can then write

$$f(z) = (z - z_0)^m r(z)$$

with $r(z_0) \neq 0$ and $r(z)$ analytic in $|z - z_0| < R_1$. Following an argument similar to that for the case (i), we know that $r(z)$ will be non-vanishing on the disc $|z - z_0| < R_2 > 0$. Again, let $R_3 = \min(R_1, R_2)$. We can now write, for $|z - z_0| < R_1$,

$$f'(z) = m(z - z_0)^{(m-1)} r(z) + (z - z_0)^m r'(z)$$

$$f''(z) = m(m-1)(z - z_0)^{(m-2)} r(z) + 2m(z - z_0)^{m-1} r'(z) + (z - z_0)^m r''(z).$$

Thus, in $|z - z_0| < R_3$, we have

$$Q(z) = -\frac{f''(z)}{f(z)}$$

$$= m(m-1)(z-z_0)^{-2} + 2m(z-z_0)^{-1}\frac{r'(z)}{r(z)} + \frac{r''(z)}{r(z)}.$$

As $r(z)$ is analytic in $|z - z_0| < R_3$, so are r' and r'', and further, as $r(z) \neq 0$ in $|z - z_0| < R_3$, $\frac{1}{r(z)}$ is also analytic there. Evidently Q has at worst a double pole at $z = z_0$. (In particular, Q has no other singularities at $z = z_0$ such as branch point, etc.)

Corollary It follows that if Q has any singularity 'worse than' a double pole (i.e., a higher order pole or an essential singularity or a singularity not of an isolated kind), then *any* solution of (1.25) *must* necessarily have a singularity at $z = z_0$.

Comments: The above result *does not necessarily imply the existence of some solution f* analytic at $z = z_0$ if Q has no worse singularity than a double pole. Thus, the converse of the theorem is not true. Later on we shall see the further conditions under which this is true.

Example Consider the ODE

$$f'' + Qf = 0$$

with (i) $Q = \frac{-2}{z^2}$ and (ii) $Q = \frac{3}{16z^2}$. In the first case $f'' - \frac{2f}{z^2} = 0$, we can easily write down two linearly independent solutions by inspection:

$$f_1 = z^2, \quad f_2 = z^{-1}.$$

Thus, in case (i) the ODE does have an analytic solution around $z = 0$. The theorem, in this case, implies that Q can have at worst a double pole at $z = 0$, which is the case. (We note that the general solution here is not analytic at $z = 0$. But the conditions of the theorem do not warrant it).

In the second case, however,

$$f'' + \frac{3}{16}\frac{f}{z^2} = 0$$

has two linearly independent solutions $f_1 = z^{1/4}$ and $f_2 = z^{3/4}$. We note that both the solutions (as well as any linear to combination of them) have a branch point at $z = 0$. This illustrates that the converse of the theorem is not true: Here, Q has (at worst) a double pole; yet there is no solution f that is analytic at $z = 0$.

The above results indicate how the analytic properties of the solution to (1.25) at $z = z_0$ is dependent on the analytic properties of p and q at $z = z_0$.

We note, in particular,

(i) Q has 'at worst a double pole' if q has 'at worst a double pole' and p has 'at worst a simple pole'.
(ii) Q has a worse singularity than a double pole only if q has a 'worse singularity' than a double pole and/or p has a 'worse singularity' than a simple pole.

In view of these results, we find the classification in the following subsection understandable.

1.4.1 Classification of points

The discussion in the previous subsection has indicated the manner in which the analytic properties of a solution to (1.25) in the neighbourhood of $z = z_0$ is dependent on those of p and Q. This discussion suggests the following classification.

1. $p(z)$ and $q(z)$ [and hence $Q(z)$] are both analytic at $z = z_0$. In this case $z = z_0$ is referred to as an *ordinary point* of the differential equation (1.25).
   In other cases, $z = z_0$ is referred to as a *singular point* of $z = z_0$. A singular point is further classified as:
2. If $p(z)$ and/or $q(z)$ is singular at $z = z_0$, and $p(z)$ has a pole at the worst of order one and $q(z)$ has a pole at the worst of order two at $z = z_0$ (and hence so does $Q(z)$: see (1.28)), then $z = z_0$ is called a regular singular point of (1.25). In this case, $A(z) \equiv (z - z_0)p(z)$ and $B(z) \equiv (z - z_0)^2 q(z)$ are both analytic at $z = z_0$.
3. If $p(z)$ and/or $q(z)$ is singular at $z = z_0$ and $p(z)$ has a pole of order higher than one and/or $q(z)$ has a pole of order greater than two then $z = z_0$ is called an *irregular singular point* of (1.25).

In this book, we will not consider the cases when $p(z)$ and/or $q(z)$ have either an essential singularity at $z = z_0$ or a non-isolated singularity at $z = z_0$. Such cases do not occur in Physics problems. (See Section 1.7 for comments to this effect.)

We will study the nature of the general solution at $z = z_0$ in each of the cases listed above.

Examples

1. Consider the equation (a) obtained in Section 1.2 for the radial coordinate. When it is written out after changing the independent variable to z and dependent variable to y, it reads

$$\frac{d^2y}{dz^2} + \frac{2}{z}\frac{dy}{dz} + \left[k^2 - \frac{l(l+1)}{z^2}\right] y = 0$$

and is called the spherical Bessel equation. Here

$$p(z) = \frac{2}{z}; \quad q(z) = k^2 - \frac{l(l+1)}{z^2}.$$

(a) Consider the point $z = 1$. At this point both p and q are analytic. Hence $z = 1$ is an *ordinary point* of the equation.
(b) $p(z)$ and $q(z)$ both have singularities only at $z = 0$. Hence $z = 0$ is a singular point of the equation. We further note that p has a pole of order one and q has a pole of order two at $z = 0$ (if $l(l+1) \neq 0$). (Equivalently, $zp(z)$ and $z^2q(z)$ are analytic at $z = 0$). Hence $z = 0$ is a regular singular point.
(c) In fact, as $p(z)$ and $q(z)$ are analytic at all points except $z = 0$, any $z \neq 0$ is an ordinary point of the equation.

2. Consider the ODE

$$\frac{d^2y}{dz^2} + \frac{1}{z(z-1)^2}\frac{dy}{dz} + \frac{2}{z(z-2)^3}y = 0.$$

Here,

$$p(z) = \frac{1}{z(z-1)^2}; \quad q(z) = \frac{1}{z(z-2)^3}.$$

(a) Both p and q are analytic in $\mathbb{C}$ except possibly at $z = 0, 1, 2$. Hence every z other than these three points is an ordinary point.
(b) $z = 0, 1, 2$ are singular points since either p or q or both are singular there. To test the nature of these singular points, we note:
   (i) $zp(z) = \frac{1}{(z-1)^2}$ and $z^2q(z) = \frac{z}{(z-2)^2}$ are both analytic at $z = 0$. Hence $z = 0$ is a regular singular point.
   (ii) $(z-1)p(z) = \frac{1}{z(z-1)}$, $(z-1)^2q(z) = \frac{(z-1)^2}{z(z-2)^2}$.
   Here, at $z = 1$, $(z-1)p(z)$ is singular [while $(z-1)^2q(z)$ is analytic.] Hence $z = 1$ is a irregular singular point.
   (iii) For $z = 2$, we have:

$$(z-2)p(z) = \frac{z-2}{z(z-1)^2}; \quad (z-2)^2q(z) = \frac{1}{z(z-2)}.$$

   $(z-2)^2q(z)$ is singular at $z = 2$. Hence $z = 2$ is also an irregular singular point.

Point at infinity

As remarked earlier, the nature of general solution to a differential equation in the neighbourhood of a point $z = z_0$ depends on the nature of the point (i.e., whether it is ordinary, regular singular or irregular singular.) For example, suppose one wants to know the behaviour of the solution to the ODE for small $|z|$. Then one would obtain the solution around $z = 0$. Then the nature of the point $z = 0$ for the ODE will allow us to know the results about the behaviour of the solution for small $|z|$.

Often, one may also be interested in knowing, in various physical contexts, the behaviour of the solutions to the ODE for large $|z|$. In such a case, then, one would like to study the nature of 'the point $z = \infty$' for a given ODE. This, then, would allow us to determine the behaviour of its solutions for large $|z|$. We, however, need to clarify what we mean mathematically by the point at infinity in this context. We shall follow the following procedure. In the ODE, we change the independent variable to $\xi = 1/z$, and transform the entire equation in terms of ξ. Now, the behaviour of the solutions of ODE for large $|z|$ is given in terms of the its behaviour for small $|\xi|$. Hence we study the transformed ODE near the point $\xi = 0$ and define the nature of point at infinity ($z = \infty$) as that of the point $\xi = 0$ for the transformed ODE.

In the following, we shall first exhibit the procedure with the help of an example and then develop the general procedure.

Example Consider the equation in the Example 1 in Section 1.4.1 (the spherical Bessel equation).

The ODE reads

$$\frac{d^2y}{dz^2} + \frac{2}{z}\frac{dy}{dz} + \left[k^2 - \frac{l(l+1)}{z^2}\right] y = 0.$$

To study the nature of the point at infinity, we change the independent variable to $\xi = 1/z$. We find

$$\frac{dy}{dz} = \frac{dy}{d\xi}\frac{d\xi}{dz} = -\frac{1}{z^2}\frac{dy}{d\xi} = -\xi^2\frac{dy}{d\xi}$$

$$\frac{d^2y}{dz^2} = \frac{d}{dz}\left(\frac{dy}{dz}\right) = \frac{d\xi}{dz}\frac{d}{d\xi}\left(-\xi^2\frac{dy}{d\xi}\right) = \xi^2\frac{d}{d\xi}\left(\xi^2\frac{dy}{d\xi}\right)$$

$$= \xi^4\frac{d^2y}{d\xi^2} + 2\xi^3\frac{dy}{d\xi}.$$

The equation then reduces to

$$\xi^4\frac{d^2y}{d\xi^2} + \left[k^2 - l(l+1)\xi^2\right] y = 0.$$

Evidently, the form of the ODE is preserved, with

$$p(\xi) = 0, \quad q(\xi) = \frac{1}{\xi^4}[k^2 - l(l+1)\xi^2].$$

As $\xi^2 q(\xi)$ is singular at $\xi = 0$, $\xi = 0$ is an irregular singular point of the above ODE. We then say that the point $z = \infty$ is an irregular singular point of the ODE.

This example illustrates the basic procedure which we now apply to the general case. Consider the ODE in the standard form

$$\frac{d^2y}{dz^2} + p(z)\frac{dy}{dz} + q(z)y = 0.$$

We make the change of variables to $\xi = \frac{1}{z}$. Substituting the results for $\frac{d^2y}{dz^2}$ obtained from the above example in the ODE,

$$\left[\xi^4\frac{d^2y}{d\xi^2} + 2\xi^3\frac{dy}{d\xi}\right] + p\left(\frac{1}{\xi}\right)(-\xi^2)\left(\frac{dy}{d\xi}\right) + q\left(\frac{1}{\xi}\right)y = 0.$$

Dividing out by ξ^4 and regrouping the terms, we find

$$\frac{d^2y}{d\xi^2} + \left[\frac{2}{\xi} - \frac{1}{\xi^2}p\left(\frac{1}{\xi}\right)\right]\frac{dy}{d\xi} + \frac{1}{\xi^4}q\left(\frac{1}{\xi}\right)y = 0.$$

We can rewrite this as

$$\frac{d^2y}{d\xi^2} + P(\xi)\frac{dy}{d\xi} + Q(\xi)y = 0.$$

Thus, the ODE preserves its general form under the change of the independent variable $z \to \xi$. We further have,

$$P(\xi) = \frac{2}{\xi} - \frac{1}{\xi^2}p\left(\frac{1}{\xi}\right); \quad Q(\xi) = \frac{1}{\xi^4}q\left(\frac{1}{\xi}\right).$$

Now, the nature of singularity at $\xi = 0$ for the transformed ODE depends on $P(\xi)$ and $Q(\xi)$. In other words, the point $\xi = 0$, or the point $z = \infty$ is

1. an ordinary point if $P(\xi)$ and $Q(\xi)$ are analytic at $\xi = 0$.
2. a regular singular point if $\xi P(\xi)$ and $\xi^2 Q(\xi)$ are both analytic at $\xi = 0$ but $P(\xi)$ and/or $Q(\xi)$ are not so.
3. an irregular singular point otherwise.

1.4.2 Commonly occurring ODE's and their properties and applications

Tables 1.1 and 1.2 give a list of the commonly occurring second order ODEs and their properties. We also point out in what type of physical problems they are relevant.

Table 1. 1

Name of the equation	Form of the equation	RS point	IRS point
SHO	$y'' + \omega^2 y = 0$	—	∞
Legendre equation	$(1 - z^2)y'' - 2zy' + l(l+1)y = 0$	$-1, 1, \infty$	—
Associated Legendre equation	$(1 - z^2)y'' - 2zy' + \left[l(l+1) - \frac{m^2}{1-z^2}\right] y = 0$	$-1, 1, \infty$	—
Laguerre equation	$zy'' + (1 - z)y' + ay = 0$	0	∞
Hermite equation	$y'' - 2xy' + 2ay = 0$	—	∞
Bessel equation	$z^2 y'' + zy' + (z^2 - \nu^2)y = 0$	0	∞
Hypergeometric equation	$z(z-1)y'' + [(a+b+1)z - c]y' + aby = 0$	0,1,∞	—
Confluent hypergeometric equation	$zy'' + (c - z)y' + ay = 0$	0	∞

Table 1. 2

Name of the equation	Useful physical context
SHO	Classical simple harmonic oscillator
Legendre	Wide class of physical problems with spherical symmetry
Associated Legendre	Wide class of physical problems with spherical symmetry
Laguerre	Hydrogen atom
Hermite	Quantum mechanical harmonic oscillator
Bessel	Electrostatics, Quantum mechanics, multipole fields
Hypergeometric	Wide class of functions can be represented in terms of its solutions
Confluent hypergeometric	Wide class of functions can be represented in terms of its solutions

1.5 Nature of the General Solution in the Neighbourhood of an Ordinary Point

Theorem 1.3

(i) There exists a unique solution to (I) with the boundary conditions $y(z_0) = a$ and $y'(z_0) = b$ which is analytic in some neighbourhood of an ordinary point $z = z_0$.

(ii) The general solution of (I) near an ordinary point $z = z_0$ is analytic at $z = z_0$.

Proof: We start from the transformed form of (I). We had defined (we set $a = z_0$ in (1.26))

$$y(z) = \exp\left[-\frac{1}{2}\int_{z_0}^{z} p(z')dz'\right] f(z), \tag{i}$$

$$Q(z) = q(z) - \frac{1}{4}p^2(z) - \frac{1}{2}p'(z), \tag{ii}$$

and obtained

$$\frac{d^2 f(z)}{dz^2} + Q(z)f(z) = 0. \tag{iii}$$

Now as $p(z)$ and $q(z)$ are analytic at $z = z_0$, so are p^2, p' and hence $Q(z)$. Thus (iii) is a simplified form of the ODE with the middle term missing and analytic properties of the coefficient functions unchanged. In other words, it is sufficient to consider ODE of the simplified form (iii) without loss of generality.

We shall prove the result by an explicit construction.

We have the differential equation

$$f''(z) + Q(z)f(z) = 0 \tag{II}$$

with the boundary conditions $f(z_0) = a$ and $f'(z_0) = b'$. We note that $Q(z) = q(z) - \frac{1}{2}p' - \frac{1}{4}p^2$ is also analytic at $z = z_0$. Thus z_0 is an ordinary point of (II).

A heuristic approach

Before we continue with the proof, we shall give a heuristic approach to the procedure used in the proof. Let us think backwards and try to understand how we would have solved (II) had we known that a solution $f(z)$, analytic at z_0 existed. We would then expand $f(z)$ in a Taylor series around z_0 in some neighbourhood

$$f(z) = \sum_{n=0}^{\infty} A_n (z - z_0)^n. \tag{i}$$

Now $f(z_0) = A_0 = a$ and $f'(z_0) = A_1 = b'$. Thus, the boundary conditions already determine the first two coefficients A_0 and A_1. The higher terms in the expansion (i) must then be determined by substituting (i) back in (I). However, unless $Q(z)$ has a particularly simple expansion [such as a monomial in $(z - z_0)$], this procedure would not help us to determine A_n's in a *simple* way and establish the existence of the solution in the form of (i) without much labour.

Instead, we devise an analogous procedure, in which, instead of a power series expansion, we look for an expansion in terms of a series of analytic functions which is a more general framework.

$$f(z) = \sum_{n=0}^{\infty} f_n(z) \tag{ii}$$

The idea is to be able to determine f_n's successively through relations that relate an f_n to only *one* earlier function f_{n-1}, making the procedure more tractable. This is in contrast to (i), where the relation determining A_n would generally involve *all* earlier A_m's $(m < n)$. The boundary conditions have already determined a part of $f(z)$, viz., $[a + b'(z - z_0)]$ in (i), which we take as $f_0(z)$. Now $f_0(z)$ satisfies the boundary condition and

$$\frac{d^2 f_0}{dz^2} \equiv 0.$$

Thus, it does not satisfy (II), but rather

$$(f_0'' + Qf_0) - Qf_0 = 0. \tag{iii}$$

We now add a new term $f_1(z)$ to f_0 such that its second derivative $\frac{d^2 f_1}{dz^2}$ now cancels the *residue* (Qf_0) from (iii). Thus, we have

$$\frac{d^2 f_1}{dz^2} - Qf_0 = 0.$$

We also choose $f_1(z_0) = f_1'(z_0) = 0$ so that $(f_0 + f_1)$ still satisfies the boundary condition. But it does not still satisfy the differential equation for

$$\frac{d^2}{dz^2}(f_0 + f_1) + Q(z)(f_0 + f_1) = 0 - Qf_0 + Qf_0 + Qf_1 = Qf_1.$$

Now, there is a new residue Qf_1, left behind. (However, as we shall later see, f_1 is, in fact, of 'higher order' than f_0 so that the new residue is 'smaller' at least in the vicinity of $z = z_0$). To cancel this, we seek to add an f_2 to $(f_0 + f_1)$ such that

$$\frac{d^2 f_2}{dz^2} + Qf_1 = 0$$

and so on. We shall see, in fact, that the successive f_n's are, like the terms in powers series (i), of a higher and higher order in $|z - z_0|$ in their *leading* behaviour and lead to a convergent series in some neighbourhood; but it is not necessarily a Taylor series.

Thus, the procedure will consist of seeking a solution as a convergent sum of analytic functions

$$f(z) = \sum_{n=0}^{\infty} f_n(z)$$

with the property

$$\frac{d^2 f_n}{dz^2} + Qf_{n-1}(z) = 0. \tag{iv}$$

As f_0 is known, one can imagine integrating these successively. What then remains to be shown is that the series converges.

Proof (contd.): We shall first outline the structure of the proof.

(i) We shall determine first the solutions to the infinite set of equations

$$\frac{d^2 f_n(z)}{dz^2} + Q(z) f_{n-1}(z) = 0 \quad n = 1, 2, \ldots$$

with $f_0(z) = a + b'(z - z_0)$ and $f_n(z_0) = f_n'(z_0) = 0$ $n = 1, 2, \ldots$.

(ii) We shall then put an upper bound on $|f_n(z)|$ with a view to establish results related to convergence.

(iii) Using this upper bound, we shall establish that $\sum_{n=0}^{\infty} f_n(z)$ in fact, converges uniformly in $|z - z_0| < R_1$, where $|z - z_0| = R_1$ is the largest circle centered at z_0 in the domain of analyticity of $Q(z)$. We call this the domain D.

(iv) Likewise, we shall prove the absolute convergence of this series in $|z - z_0| < R_1$.

(v) We can then add the equations of the form (i) and have

$$\sum_{n=1}^{\infty} \left[\frac{d^2 f_n}{dz^2} + Q(z) f_{n-1}(z) \right] = 0$$

and, on account of the absolute convergence of $\sum f_n$ and $\sum \frac{d^2 f_n}{dz^2}$, we can write

$$\sum_{n=1}^{\infty} \frac{d^2 f_n}{dz^2} + Q(z) \sum_{n=1}^{\infty} f_{n-1}(z) = 0.$$

(vi) On account of uniform convergence of $\sum_n f_n$, we can write

$$\frac{d^2}{dz^2} \sum_{n=0}^{\infty} f_n(z) = \frac{d^2}{dz^2} \sum_{n=1}^{\infty} f_n(z) = \sum_{n=1}^{\infty} \frac{d^2 f_n(z)}{dz^2} = \frac{d^2 f}{dz^2}.$$

(vii) This then will complete the proof of existence of an analytic solution $f(z) = \sum_{n=0}^{\infty} f_n(z)$ at an ordinary point $z = z_0$ with specified boundary conditions.

Now we come to the details of the proof outlined above.

(i) We let $f(z)$ as given by (ii). We construct f_n by direct integration of (iv) in terms of f_{n-1}. We note that

$$\frac{df_n}{dz} = -\int_{z_0}^{z} Q(z') f_{n-1}(z') dz' \tag{v}$$

satisfies the boundary condition $f'_n(z_0) = 0$. We further verify that

$$f_n(z) = -\int_{z_0}^{z} (z - z')Q(z')f_{n-1}(z')dz' \tag{vi}$$

satisfies the additional boundary condition $f_n(z_0) = 0$ and also leads to (v) upon explicit differentiation.

The integrals in (v) and (vi) are taken over an arbitrary path lying in D. For convenience of computation and estimation of $|f_n(z)|$, it is convenient to choose a straight line segment from z_0 to z. We parameterize the straight line by a parameter $0 \le t \le 1$:

$$z' = z_0 + t(z - z_0).$$

Then,

$$dz' = (z - z_0)dt$$

and

$$(z - z') = (z - z_0)(1 - t)$$

and

$$f_n(z) = -(z - z_0)^2 \int_0^1 dt(1 - t)Q(z'(t))f_{n-1}(z'(t)).$$

(ii) The above relation indicates that there are two extra powers of $(z - z_0)$ between any two successive f_n's. We would, thus, expect $f_n(z) = O[(z - z_0)^{2n}]$ in the vicinity of $z = z_0$. We further see that the above expression for f_n is directly proportional to Q. If thus Q is scaled by λ, we expect f_n to be scaled by λ^n. These considerations allow us to conjecture

$$|f_n(z)| \le \frac{|z - z_0|^{2n} Q_0^n |f_0|_{\max}}{n!}, \tag{1.29}$$

where Q_0 = maximum value of $|Q(z'(t)|$ for $t \in [0, 1]$,

$|f_0|_{\max}$ = maximum value of $f_0(z'(t))$ $t \in [0, 1]$.

In fact, we shall prove this result by induction. To begin, we shall assume that for any $z'(t), 0 \le t \le 1$,

$$|f_{n-1}(z')|)| \le \frac{|z' - z_0|^{2n-2} Q_0^{n-1} |f_0|_{\max}}{(n-1)!} = \frac{t^{2n-2}|z - z_0|^{2n-2} Q_0^{n-1} |f_0|_{\max}}{(n-1)!}.$$

Then,

$$|f_n(z)| = |z - z_0|^2 \left| \int_0^1 dt(1-t)Q(z'(t))f_{n-1}(z'(t)) \right|$$

$$\leq |z - z_0|^2 \int_0^1 dt(1-t)|Q(z'(t))||f_{n-1}(z'(t))|$$

$$\leq |z - z_0|^{2n} \int_0^1 dt(1-t)Q_0(1-t)^{2n-2}\frac{Q_0^{n-1}}{(n-1)!}|f_0|_{\max}$$

$$= \frac{|z - z_0|^{2n}Q_0^n|f_0|_{\max}}{(n-1)!}\int_0^1 dt(1-t)^{2n-1}$$

$$= \frac{|z - z_0|^{2n}Q_0^n|f_0|_{\max}}{(n-1)!}\frac{1}{2n}$$

$$< \frac{|z - z_0|^{2n}Q_0^n|f_0|_{\max}}{n!} \qquad n = 1, 2, \ldots$$

Further, the relation holds trivially for $n = 0$. This completes the proof by induction.

(iii) Now,

$$\sum_{n=0}^{\infty}|f_n(z)| \leq \sum \frac{|z - z_0|^{2n}Q_0^n}{n!}|f_0|_{\max} = |f_0|_{\max}(\exp Q_0|z - z_0|^2) < \infty. \tag{1.30}$$

(iv) Thus, the series $\sum f_n(z)$ is absolutely convergent for all z. Moreover, consider a z such that $|z - z_0| < R < R_1$.

Consider the sum

$$\left|\sum_{n=N}^{\infty} f_n(z)\right| \leq \sum_{n=N}^{\infty}|f_n(z)| \leq \sum_{n=N}^{\infty}|f_0|_{\max}\frac{|Q_0(z - z_0)^2|^n}{n!}$$

$$\leq \sum_{n=N}^{\infty}|f_0|_{\max}\frac{(Q_0R^2)^n}{n!}.$$

Now, since the series $\sum_{n=N}^{\infty} |f_0|_{\max} \frac{(Q_0 R^2)^n}{n!}$ is convergent, we know that there exists N_0 such that

$$\left| \sum_{n=N}^{\infty} (f_0)_{\max} \frac{(Q_0 R^2)^n}{n!} \right| < \epsilon \text{ whenever } N > N_0. \text{ Thus,}$$

$$\left| \sum_{n=N}^{\infty} f_n(z) \right| < \epsilon \quad \text{for} \quad N > N_0,$$

N_0 being independent of z for any z in the disc $|z - z_0| < R$. This proves the uniform convergence of the series $\sum_{n=0}^{\infty} f_n(z)$ in D.

We thus have completed the steps outlined in the proof. The above results show that there is an analytic solution corresponding to a set of given boundary conditions of $y(z)$: $y(z_0) = a$, $y'(z_0) = b$. Thus, there is an analytic solution for each of the boundary conditions:

$$y_1(z_0) = 1 \qquad y_1'(z_0) = 0 \tag{1.31}$$

$$y_2(z_0) = 0 \qquad y_2'(z_0) = 1 \tag{1.32}$$

As is easy to see, they cannot be proportional and hence are linearly independent, and thus constitute a complete set of linearly independent solutions. Hence, the most general solution to the ODE around $z = z_0$,

$$y = ay_1 + by_2$$

is also analytic at $z = z_0$.

Comments: As is seen easily, the solution to the ODE satisfying the boundary conditions $y(z_0) = a$ and $y'(z_0) = b$ is $y = ay_1 + by_2$.

Solved problem 1.9 Consider the Hermite differential equation

$$y'' - 2zy' + 2\alpha y = 0. \tag{i}$$

Cast the differential equation in the form of

$$f'' + Qf = 0. \tag{ii}$$

Suppose we want to solve (i) with the boundary condition $y(0) = 1$, $y'(0) = 0$. Obtain the boundary condition of f in (ii). Then,

1. Write down the series of differential equations satisfied by f_n.
2. Write down $f_0(z)$ satisfying these boundary conditions.

3. Solve explicitly for f_n's for the case $\alpha = -\frac{1}{2}$.
4. Verify the bound in (1.29) and convergence properties of $\sum f_n(z)$.

Solution: Here, $p(z) = -2z$.

We make the change of the dependent variable

$$y(z) = \exp\left[-\frac{1}{2}\int^z dz' p(z')\right] f = f(z)e^{z^2/2}.$$

$$y = e^{z^2/2} f$$

$$y' = ze^{z^2/2} f + e^{z^2/2} f'$$

$$y'' = e^{z^2/2} f'' + 2zf' e^{z^2/2} + e^{z^2/2} z^2 f + e^{z^2/2} f$$

Thus,

$$y'' - 2zy' + 2\alpha y = e^{z^2/2}[f'' + (2\alpha + 1 - z^2)] = 0.$$

We thus have

$$f'' + (2\alpha + 1 - z^2) f = 0.$$

Alternately, noting $p = -2z$ and $q = 2\alpha$, we have,

$$Q(z) = q - \frac{1}{2}p' - \frac{p^2}{4} = 2\alpha + 1 - z^2.$$

1. The boundary conditions on f are

$$y(0) = f(0) = 1 = a'; \quad y'(0) = f'(0) = 0 = b'.$$

2. This leads to

$$f_0(z) = a' + b'z = 1.$$

3. The differential equation connecting successive f_n's is

$$\frac{d^2 f_n}{dz^2} + (2\alpha + 1 - z^2) f_{n-1} = 0 \qquad n = 1, 2, 3, \ldots$$

For $\alpha = -\frac{1}{2}$, it reduces to

$$\frac{d^2 f_n}{dz^2} - z^2 f_{n-1} = 0.$$

Using $f_0(z)$, we find that f_1 satisfies the equation $\frac{d^2 f_1}{dz^2} = z^2$ with $f_1(0) = f_1'(0) = 0$.
This implies that $f_1 = \frac{z^4}{12}$.
Similarly,

$$\frac{d^2 f_2}{dz^2} = z^2 f_1 \Rightarrow f_2 = \frac{z^8}{3 \cdot 4 \cdot 7 \cdot 8}.$$

In general,

$$f_n = \frac{z^{4n}}{(3 \cdot 4) \times (7 \cdot 8) \times (11 \times 12) \times \cdots \times (4n-1)(4n)}.$$

4. The bound in the proof reads

$$|f_n(z)| \leq \frac{|z - z_0|^{2n} Q_0^n |f_0|^n}{n!}.$$

Here, $z_0 = 0, \quad Q(z) = -z^2 \quad f_0 = 1$.

$Q(z)$ is entire. We shall verify the bound in the unit circle (this could however be done for *any* circle $|z| < R$)).

In this circle $|Q(z)| = |z^2| < 1 \equiv Q_0$.
Thus, for $|z| < 1$, the bound reads

$$|f_n(z)| \leq \frac{|z|^2 n}{n!}.$$

We verify this bound as follows:

$$|f_n(z)| = \frac{|z|^{4n}}{3.4 \times 7.8 \times \cdots \times (4n-1).4n}$$

$$= \frac{|z|^{2n}}{n!} |z|^{2n} \frac{1}{3.4} \times \frac{2}{7 \times 8} \times \cdots \times \frac{n}{(4n-1).4n}$$

$$< \frac{|z|^{2n}}{n!} \quad \text{for} \quad |z| < 1.$$

This verifies the bound.

Comment: The above example was given as a means towards attaining comprehension of the steps in the general proof of the above theorem. The proof of the theorem applies to a very general analytic $Q(z)$. In the case of the particularly simple $Q(z) = -z^2$, the simple method of series solution (alluded to in the heuristic approach) works in a tractable manner. This is worked out below.

Solved problem 1.10 Solve the Hermite differential equation

$$y'' - 2zy' + 2\alpha y = 0$$

for $\alpha = -1/2$ around $z = 0$ with boundary conditions $y(0) = 1$, $y'(0) = 0$.

Solution: Since $z = 0$ is an *ordinary* point of the differential equation, a solution analytic at $z = 0$ exists. Hence, it can be expanded as a Taylor series.

We write

$$y(z) = \sum_{n=0}^{\infty} a_n z^n$$

and note that the boundary conditions imply that

$$y(0) = a_0 = 1; \qquad y'(0) = a_1 = 0.$$

We substitute the series in the differential equation. We have, for $\alpha = -\frac{1}{2}$,

$$y'' - 2zy' - y = 0,$$

$$\sum_{n=0}^{\infty} a_n n(n-1)z^{n-2} - 2\sum_{n=0}^{\infty} a_n n z^n - \sum_{n=0}^{\infty} a_n z^n = 0.$$

We compare the coefficients of each power $z^n\,(n \geq 2)$. We then have

$$a_{n+2}(n+2)(n+1) = (2n+1)a_n.$$

Thus,

$$a_{n+2} = a_n \frac{(2n+1)}{(n+1)(n+2)} \qquad n \geq 0.$$

We note that the above 'recurrence relations' determines all a_n's in terms of a_0 and a_1. The latter can be known from the boundary condition. We note

$$y(0) = a_0 = 1; \ldots y'(0) = a_1 = 0.$$

Thus, all the 'odd coefficients' a_{2m+1} vanish. The even coefficients are then obtained using

$$a_{2m+2} = a_{2m} \frac{(4m+1)}{(2m+1)(2m+2)},$$

i.e.,

$$a_{2m} = \frac{(4m-3)}{(2m)(2m-1)} \times \frac{(4m-7)}{(2m-2)(2m-3)} \times \cdots \times \frac{(-3)}{2 \times 1}$$

$$= \frac{(4m-3)(4m-7)\ldots(-3)}{(2m)!}.$$

1.6 Solution in the Vicinity of a Regular Singular Point

We shall now discuss the solution to the second order homogeneous differential equation

$$y''(z) + p(z)y'(z) + q(z)y(z) = 0 \tag{1.33}$$

in the vicinity of a regular singular point. We shall demonstrate that in the vicinity of a regular singular point, $z = z_0$, *at least one* solution of (1.33) can be obtained in the series form

$$y(z) = (z - z_0)^r \sum_{n=0}^{\infty} c_n (z - z_0)^n. \tag{1.34}$$

We shall also show that for a large class of differential equations, a second linearly independent solution of a *similar form* exists and can be found simultaneously by what is called the *series solution method*. We shall develop the criteria when this is possible. In those cases, where this is not possible, we always have a recourse to the method already outlined earlier; viz., the Wronskian method to obtain a second linearly independent solution. Using this we shall establish the *form* of the second linearly independent solution. We shall show that this second solution, apart from a series of the form (1.34), generally contains a term proportional to $\ln(z - z_0) \times y_1$; y_1 being the first solution obtained by the series method employing (1.34). Using these general results, we shall be able to show that unlike the solution in the vicinity of an ordinary point, the *general* solution will, *barring exceptional cases*, exhibit the presence of a singularity at $z = z_0$.

1.6.1 Series solution method: general dicussion

In this section, we shall develop the method of series solutions. We shall first give a broad outline of its development. The development of the method proceeds as follows:

- Assume that a solution in the form of a convergent series (1.34) exists and substitute it in (1.33).
- Arrive at the (infinite number of) conditions that must be satisfied by the set of coefficients $(c_0, c_1, c_2, \ldots\ldots c_n, \ldots\ldots)$ obtained by equating the coefficients of various powers of z in (1.33).
- Show that a solution of this form can exist (if at all) for only two values of r, viz., r_1 and r_2, determined by what is called the 'indicial equation'.
- Show that, for a certain of these two roots, $r = r_1$, the infinite set of conditions on $(c_0, c_1, c_2, \ldots\ldots c_n, \ldots\ldots)$ have a *unique* solution for all the ratios $\frac{c_n}{c_0}$.
- Show further that, with this set of values of $\frac{c_n}{c_0}$; $n = 1, 2, 3, \ldots\ldots$ the series in (1.34) is in fact convergent in *some* neighbourhood $|z - z_0| < R$ (and therefore *uniformly convergent* there), justifying the original assumptions and the procedure that uses term-by-term differentiation.

- Show further that $r = r_2$ (i.e., the *other* root) also yields a second *linearly independent solution* in the form of the series (1.34) *provided* $r_1 - r_2 \neq$ *an integer.*

Having given an outline of the method, we would like to start with the proof. For simplicity, and without loss of generality, we shall set the point $z_0 = 0$. We multiply (1.33) by z^2 and write it as

$$z^2 y''(z) + A(z)y'(z) + B(z)y(z) = 0 \tag{1.35}$$

with

$$A(z) \equiv zp(z) \quad B(z) \equiv z^2 q(z),$$

both of which are known to be analytic at the regular singular point $z = 0$. In view of this, there exists a Taylor expansion for $A(z)$ and $B(z)$ valid within circles of radii R_1 and R_2 respectively. Thus, for $|z - z_0| < R_3 \equiv \min(R_1, R_2)$, we have

$$A(z) = \sum_{s=0}^{\infty} a_s z^s; \quad B(z) = \sum_{s=0}^{\infty} b_s z^s. \tag{1.36}$$

We now substitute the series (1.34) in (1.33). We note that we have *assumed* that the series converges in some domain $|z| < R$ and consequently is also uniformly convergent there. This allows us to carry out term-by-term differentiation of (1.34). We shall then equate the coefficients of various powers of z. The lowest power of z that appears in (1.33) is z^r. Its coefficient is (as seen by inspection):

$$c_0\{r(r-1) + a_0 r + b_0\} = 0.$$

We can always choose, without loss of generality, $c_0 \neq 0$ in (1.34). This implies that the solution of the form (1.34) is possible only if r satisfies what is called the 'indicial equation'

$$r^2 + (a_0 - 1)r + b_0 = 0$$

the two roots of which are

$$r_{1,2} = \frac{(1 - a_0) \pm \sqrt{(a_0 - 1)^2 - 4b_0}}{2}.$$

We note that, in the general case, the two roots $r_{1,2}$ can be complex. We shall by convention choose to designate by r_1 that root for which the real part is not lesser; i.e., $\mathrm{Re}\, r_1 \geq \mathrm{Re}\, r_2$. (Note that if $\mathrm{Re}\, r_1 = \mathrm{Re}\, r_2$, then it does not matter which is called which). The presence of the indicial equation restricts the leading power in z in a possible series solution of the form (1.34).

To proceed, we first note the form of the various terms on the left-hand side of (1.35):

$$z^2 y'' = \sum_{n=0}^{\infty} c_n (n+r)(n+r-1) z^{n+r}$$

$$A(z)y(z) = \sum_{s=0}^{\infty} a_s z^s \sum_{n=0}^{\infty} c_n (n+r) z^{n+r}$$

$$B(z)y(z) = \sum_{s=0}^{\infty} b_s z^s \sum_{n=0}^{\infty} c_n z^{n+r} \tag{1.37}$$

and make several observations on the coefficients of z^{r+m} for $m > 0$:

- An inspection of the terms on the left-hand side of (1.35) as given above in (1.37) will show that the 'highest coefficient' c_n that enters into the coefficient of z^{m+r} is for $n = m$. Thus equating this coefficient of z^{m+r} to zero will lead to an equation that contains in it at most the coefficients $(c_0, c_1, c_2, \ldots\ldots c_m)$. This is true successively as m increases: $m = 1, 2, 3, \ldots$.
- All these equations are *linear homogeneous* in the c_n's (see the form of the right-hand side of (1.37)). Equivalently, they depend linearly only on the *ratios* $\frac{c_n}{c_0}$.
- In such an equation, this highest coefficient c_m is multiplied by

$$= [(m+r)(m+r-1) + a_0(m+r) + b_0]$$
$$= [(m+r)^2 + (a_0 - 1)(m+r) + b_0]. \tag{1.38}$$

- Whether we can obtain the values of the 'new coefficient' c_m from this (mth equation) depends crucially on whether this coefficient $[(m+r)^2 + (a_0 - 1)(m+r) + b_0]$ is *non-vanishing*.
- This factor can *vanish* only if $(m+r) = \frac{(1-a_0)\pm\sqrt{(a_0-1)^2-4b_0}}{2}$. We note that the right-hand side has just the roots r_1 and r_2.
- Now when we are attempting the series solution for $r = r_1$, the above condition (for vanishing) requires $(m + r_1) = r_1$ or $(m + r_1) = r_2$. The first is not possible for any $m > 0$. The second also is impossible because $\mathrm{Re}(m + r_1) > \mathrm{Re}(r_1) \geq \mathrm{Re}(r_2)$. Thus, $(m + r_1)$ can never equal r_1 or r_2 for any $m > 0$. Thus, the factor in (1.38) does not vanish when $r = r_1$ for any $m > 0$.
- When we are attempting a series solution for $r = r_2$, $(m + r_2)$ cannot be r_2 for any $m > 0$. $(m + r_2)$ *can* be r_1 however *only if* $r_1 - r_2 = m$ is a positive integer.

With these observations, we shall now show that for the case $r = r_1$, the equations obtained by equating successively the higher and the higher powers z^{m+r} to zero have a unique solution for all the ratios $\frac{c_n}{c_0}$. We proceed via induction. Let the equations obtained by equating to zero the coefficients of $z^r, z^{r+1}, \ldots z^{r+p}$ be solved and let this have yielded a unique solution for $\frac{c_n}{c_0}, n = 1, 2, \ldots, p$. Let us now consider the equation obtained by equating the coefficient of z^{r+p+1} to zero. The *only* unknown that this coefficient will contain is c_{p+1} and hence this additional equation will have the form

$$[(p+1+r)^2 + (a_0 - 1)(p+1+r) + b_0]\frac{c_{p+1}}{c_0} = \text{An already determined quantity.}$$

Thus, $\frac{c_{p+1}}{c_0}$ is uniquely determined *provided*

$$[(p+1+r)^2 + (a_0 - 1)(p+1+r) + b_0] \neq 0.$$

Now, for $r = r_1$, we have shown that this coefficient does not vanish. (It is in fact seen to be $(p+1)(p+1+r_1-r_2) \neq 0$ after some simplification.) Hence, these equations can be solved for $\frac{c_n}{c_0}$; $n = 1, 2, 3, \ldots, (p+1)$. A similar result follows also for the second root r_2 when $r_1 - r_2 \neq$ integer. We shall soon show that the series so obtained is convergent in some domain $|z| < R' > 0$. It thus follows that a series of the form (1.34) with $r = r_1$ always exists. When $r_1 = r_2$, we cannot have another series as the there is only one allowed value for the index r.

When $r_1 - r_2 \neq$ integer, we have another series solution of the form (1.34). This solution is linearly independent of the first one as $r_1 \neq r_2$.

Proof of convergence: We shall now briefly sketch the proof of the convergence of the series (1.34) for $r = r_1$. To this end, we shall first prove a bound on the coefficients c_n's. We shall do this by induction. For this purpose we will need the exact expression for c_n. It is given by

$$c_n = \frac{\sum_{r=1}^{n-1} c_{n-r}[(r_1 + n - r)a_r + b_r] + r_1 a_n + b_n}{n(n + r_1 - r_2)}. \tag{1.39}$$

We refer to the equation (1.36). We note that the expansion coefficients a_n and b_n in the Taylor expansion of $A(z)$ and $B(z)$ are related to the respective derivatives which obey the bounds arising from the Cauchy integral representations. There exists a positive integer M independent of n such that

$$|a_n| < M\rho^n, \qquad |b_n| < M\rho^n, \qquad |r_1 a_n + b_n| < M\rho^n. \tag{1.40}$$

Without loss of generality, we can assume $M > 1$. We note that

$$c_1 = \frac{r_1 a_1 + b_1}{(r+1)} < \frac{M}{r}.$$

We shall now proceed by induction. Let $c_m < M^m \rho^{-m}$ for $m = 1, 2, \ldots, n-1$. Using the above expression for c_n, it is not difficult to show that

$$|c_n| < M^n \rho^{-n},$$

thus proving the result for $m = n$. Given this bound, it follows that the series (1.34) is convergent in $|z - z_0| < \frac{\rho}{M}$.

Comments: We add a number of comments.

(a) The fact that (I) can have a *series* solution of the form (1.34) is liable to be confused with an *analytic* solution as around an ordinary point. The series solution (1.34) represents an analytic solution in the vicinity of $z = z_0$ only when the index r is a non-negative integer. When r is a fraction, the series has a branch-point at $z = z_0$ and when r is a negative integer, it has a pole.

(b) We note that r_1 and r_2 can, in principle, be complex. In that case, the solutions will necessarily be multivalued. We do not expect a physically relevant solution to be multivalued. A study of ODEs commonly occurring in Physics shows that this does not happen (see Exercise 1.23).

Solved problem 1.11 Obtain a set of two linearly independent solutions of the Bessel equation around $z = 0$.

$$y'' + \frac{1}{z}y' + \left(1 - \frac{\nu^2}{z^2}\right) y = 0 \tag{1.41}$$

when (ν is real) $2\nu \neq$ integer.

Solution: We note that here $p(z) = \frac{1}{z}$ and $q(z) = 1 - \frac{\nu^2}{z^2}$. Thus, $z = 0$ is a regular singular point since $A(z) = zp(z) = 1$ and $B(z) = z^2 q(z) = (z^2 - \nu^2)$ are analytic at $z = 0$. Now, a_0 and b_0 occurring in the indicial equation are given by: $a_0 = A(0) = 1$ and $b_0 = B(0) = -\nu^2$. Thus, the indicial equation reads

$$r(r - 1) + r - \nu^2 = r^2 - \nu^2 = 0.$$

The two roots of the indicial equation are $r = \pm\nu$. Without loss of generality, we can let $\nu > 0$. (Recall that $2\nu \neq 0$). Then $r_1 = \nu$, $r_2 = -\nu$. As $2\nu \neq$ integer, the difference of the two roots of the indicial equation is not an integer. Hence, in the present case, the two linearly independent solutions can be chosen to have the form

$$y_1 = z^\nu \sum_{n=0}^{\infty} c_n z^n,$$

$$y_2 = z^{-\nu} \sum d_n z^n.$$

Both of these solutions can be simultaneously found if we substitute

$$y = z^r \sum c_n z^n$$

in the ODE and we keep r arbitrary for the present. This leads us to

$$\sum_{n=0}^{\infty} c_n(n+r)(n+r-1)z^{n+r-2} + \sum_{n=0}^{\infty} c_n(n+r)z^{n+r-2}$$
$$+\sum_{n=0}^{\infty} c_n z^{n+r} - \nu^2 \sum c_n(n+r)z^{n+r-2} = 0.$$

We now compare the coefficient of z^{r-2} to obtain

$$c_0[r(r-1)+r-\nu^2] = c_0(r^2-\nu^2) = 0.$$

Assuming $c_0 \neq 0$, this implies $r^2 = \nu^2$.

This is just the indicial equation, as expected. Further, comparing the coefficient of z^{r-1} gives

$$c_1[(r+1)r+(r+1)-\nu^2] = c_1[(r+1)^2-\nu^2] = 0.$$

Now since $\nu^2 = r^2$, $(r+1)^2 = \nu^2$ is also satisfied only if $r = -\frac{1}{2}$. This necessarily requires that $\nu = \frac{1}{2}$ and thus $2\nu = 1$, an integer. In this problem, we are excluding such a possibility. Hence $c_1 = 0$.

Now, comparing the coefficient of z^{n+r}, $n \geq 0$, gives

$$c_{n+2}[(n+r+2)(n+r+1)+(n+r+2)-\nu^2]+c_n = 0.$$

i.e.,

$$c_{n+2} = -c_n \frac{1}{(n+r+2)^2-\nu^2} = -\frac{c_n}{(n+r+\nu+2)(n+r-\nu+2)}.$$

We note again that the denominator cannot vanish unless ν is an integer (which it is not).

We can now solve for c_n and obtain

$$c_{2m+1} = 0 \quad m = 0, 1, 2, \ldots.$$

i.e., all 'odd coefficients' vanish, and for $r = r_1 = \nu$,

$$c_{2m} = -\frac{(-)^m c_0}{(2m+2\nu)(2m+2\nu-2)\ldots(2\nu+2)\times(2m)(2m-2)\ldots(2)}$$
$$= \frac{(-)^m c_0 \nu!}{2^{2m}(m+\nu)!m!},$$

while for $r = r_2 = -\nu$,

$$c_{2m} = -\frac{(-)^m c_0}{2^{2m} m!} \frac{(-\nu)!}{(m-\nu)!}$$

The choice of c_0 is arbitrary. It is conventional to choose[8]

$$c_0 = \frac{1}{(\pm\nu)!} \cdot \frac{1}{2^{\pm\nu}}$$

for $r = \pm\nu$. Then the two linearly independent solutions are

$$J_\nu(z) = \sum_{m=0}^{\infty} \frac{(-)^m}{(m+\nu)!m!} \left(\frac{z}{2}\right)^{2m+\nu} \tag{1.42}$$

and

$$J_{-\nu}(z) = \sum_{m=0}^{\infty} \frac{(-)^m}{(m-\nu)!m!} \left(\frac{z}{2}\right)^{2m-\nu}$$
$$= J_\nu(z)|_{\nu\to-\nu}.$$

Comments:

(a) The above expression on the right-hand side of (1.42) is in fact the definition of Bessel function of order ν. We shall study these extensively in Chapter 3.

(b) We note that the two solutions J_ν and $J_{-\nu}$ are linearly independent when $\nu \neq$ an integer. This can be seen by looking at their leading behaviour ($\sim z^{\pm\nu}$) near $z = 0$.

(c) We note that the entire procedure above goes through even for $\nu =$ half odd integer and yields *both* the linearly independent solutions. In this case $r_1 - r_2 = 2\nu$ is an integer; even then the above procedure yields *two* linearly independent solutions of the series form. These are, however, exceptional cases. Towards the end of Section 1.6.2, we shall understand when this happens.

1.6.2 Second linearly independent solution by the Wronskian method

In the case when the two roots of the indicial equation r_1 and r_2 differ by an integer, we have to develop a different method for the construction of a second linearly independent solution. We already know that given one solution of the ODE, we can construct a second linearly independent solution y_2 by the Wronskian method outlined in Section 1.3.4. We shall employ the expression for y_2 in terms of y_1 given by (1.24), viz.,

$$y_2(z) = y_1(z) \int^z \frac{dz'}{y_1(z')^2} \exp\left[-\int^{z'} p(z'')dz''\right] \tag{1.43}$$

[8] For the definition of $(\pm\nu)!$ when ν is not a non-negative integer, see the succeeding chapter.

and the already-determined form of the first solution y_1:

$$y_1(z) = (z - z_0)^{r_1} \sum_{n=0}^{\infty} c_n (z - z_0)^n. \tag{1.44}$$

We shall simplify the discussion by setting $z_0 = 0$. Since $p(z)$ has (at worst) a simple pole at $z = z_0$ with residue a_0, we can write

$$p(z) = \frac{a_0}{z} + p_0(z),$$

where $p_0(z)$ is analytic in some neighbourhood $|z| < R_1$ of $z = z_0 = 0$. Then,

$$\exp\left[-\int^{z'} p(z'')dz''\right] = \exp\left[-a_0 \ln z' - \int^{z'} dz'' p_0(z'')\right] \tag{1.45}$$

$$= z'^{-a_0} \exp\left[-\int^{z'} dz'' p_0(z'')\right] \tag{1.46}$$

$$\equiv \frac{1}{z'^{a_0}} \tilde{A}_1(z'). \tag{1.47}$$

Here, $\tilde{A}_1(z)$ is analytic in $|z| < R_1$, being the exponential of an analytic function $[-\int^{z'} dz'' p_0(z'')]$.

We can further write,

$$y_1 = z^{r_1} \sum_{n=0}^{\infty} c_n z^n$$

$$= z^{r_1} \tilde{A}_2(z). \tag{1.48}$$

The *power* series in the expression for y_1 is a convergent series in some neighbourhood $|z| < R_2$ and thus represents an analytic function $\tilde{A}_2(z)$ there. Moreover, this analytic function does not vanish at $z = 0$ since $\tilde{A}_2(0) = c_0 \neq 0$. Hence, its nearest zero is at a finite distance from the origin, say at $z = R_3 \exp(i\phi)$. Then, $\frac{1}{[\tilde{A}_2(z)]^2}$ is analytic in $|z| < \min[R_2, R_3]$. Thus, $y_2(z)$ is given by

$$y_2(z) = y_1(z) \int^z \frac{dz'}{z'^{2r_1 + a_0}} \frac{\tilde{A}_1(z')}{[\tilde{A}_2(z')]^2}. \tag{1.49}$$

Now, $\tilde{A}(z) = \frac{\tilde{A}_1(z')}{[\tilde{A}_2(z')]^2}$ is analytic in $|z| < R \equiv \min[R_1, R_1, R_3]$ and consequently we can expand it in a Taylor series

$$\tilde{A}(z) = \sum_0^{\infty} d_n z^n$$

with $\tilde{A}(0) = d_0 \neq 0$. Also, we note that $2r_1 + a_0 = r_1 - r_2 + 1$. Thus, the expression for y_2 becomes

$$y_2(z) = y_1(z) \int^z \frac{dz'}{z'^{r_1 - r_2 + 1}} \sum_0^\infty d_n z'^n$$

$$= y_1(z) \int^z dz' \sum_0^\infty d_n z'^{n - r_1 + r_2 - 1}.$$

We note:

(a) We can integrate the above series term by term for any $|z| < R$, being a convergent series in this domain. The result converges in $|z| < R$.

(b) For $n = r_1 - r_2 =$ (a non-negative integer), the integral of the term gives a logarithm. This leads to a term proportional to $\ln z \times y_1(z)$ in the expression for y_2 *provided* the coefficient d_n for $n = r_1 - r_2$ (denoted henceforth by $\tilde{D}$) is nonzero.

(c) The lowest power of z in the integral is $z^{-r_1+r_2}$ when $(r_1 - r_2) \neq 0$ and z^1 in the latter case. The right-hand side of the above equation, apart from the $\ln z \times y_1(z)$ term, has $[y_1(z) \times$ (the series)$]$. The lowest power of z in it is z^{r_2} when $(r_1 - r_2) \neq 0$ and $z^{(r_2+1)}$ when $r_1 = r_2$. $[y_1(z) \times$ (the series)$]$ has the form

$$[y_1(z) \times \text{(the series)}] = z^{r_2} \tilde{A}_4(z) \tilde{A}_1(z).$$

The function $\tilde{A}_4(z)\tilde{A}_1(z)$ is analytic in $|z| < R$ and can be Taylor-expanded there.

Thus, we obtain the general form for y_2:

$$y_2(z) = \tilde{D} \ln z \times y_1(z) + z^{r_2} \sum_{p \neq r_1 - r_2}^{\infty} g_p z^p. \tag{1.50}$$

The general solution and its analytic properties

The general solution for the ODE is given by

$$y(z) = a y_1(z) + b y_2(z) \qquad a, b \in \mathbb{C}.$$

In the case when $r_1 - r_2$ is not an integer, the form of the two solutions is given by

$$y_1(z) = (z - z_0)^{r_1} \sum_{p=0}^{\infty} c_p (z - z_0)^p,$$

$$y_2(z) = (z - z_0)^{r_2} \sum_{p=0}^{\infty} d_p (z - z_0)^p. \tag{1.51}$$

In the case when $r_1 - r_2$ *is* an integer, the form of the two solutions is given by

$$y_1(z) = (z - z_0)^{r_1} \sum_{p=0}^{\infty} c_p (z - z_0)^p,$$

$$y_2(z) = \tilde{D} \ln(z - z_0) \times y_1(z) + (z - z_0)^{r_2} \sum_{p=0}^{\infty} d_p (z - z_0)^p. \tag{1.52}$$

We note:

(a) The presence of any singularity in y_1 or y_2 or both reflects in the general solution y.
(b) When r_1 (or r_2) is not an integer, the corresponding solution has a branch-cut at $z = z_0$.
(c) When r_1 (or r_2) is a negative integer, the corresponding solution has a pole at $z = z_0$.
(d) Unless $(r_1 - r_2)$ is an integer, at least one of them leads to a branch-cut in y at $z = z_0$.
(e) When $(r_1 - r_2)$ is an integer, the presence of a $\ln z \times y_1$ term implies a branch-cut at $z = z_0$ in y_2 unless $\tilde{D} = 0$.
(f) Thus, both the solutions cannot be analytic at $z = z_0$ unless

1. both r_1 and r_2 are non-negative *integers* (which requires that $r_1 - r_2 =$ an integer), and
2. the coefficient $\tilde{D}$ of the $\ln z \times y_1$ term in y_2, (Note: $r_1 - r_2 =$ integer here) vanishes fortuitously.

(g) In the special case of $r_1 = r_2$, we note that $\tilde{D} \neq 0$. Hence, in this case the $\ln z \times y_1$ term is *necessarily* present.

Hence, we find that the general solution in the vicinity of a regular singular point $z = z_0$ *barring exceptional cases* has a singularity at $z = z_0$. The exceptional cases can be located from the comment (f) above. Also see Exercise 1.41 for an example and Exercise 1.42.

In the example below, we shall see illustration of the comment (g). An illustration of the comment (e) is found in the previous solved example.

Solved problem 1.12 Obtain a set of two linearly independent solutions to the Bessel equation (for $\nu = 0$) around $z = 0$.

$$y'' + \frac{1}{z} y' + y = 0$$

Solution: As noted earlier, $z = 0$ is a regular singular point of the ODE. The indicial equation is

$$r^2 - \nu^2 = r^2 = 0.$$

The two roots are $r_1 = 0 = r_2$. Thus, this is the case of two *equal roots* for the indicical equation and thus the two solutions have the form

$$y_1 = \sum_{n=0}^{\infty} c_n z^n \tag{i}$$

and

$$y_2 = \ln z \times y_1 + \sum_{n=1}^{\infty} d_n z^n \equiv y_0 + \tilde{y}_2. \tag{ii}$$

As the two indices are equal, the logarithmic term in y_2 cannot be avoided.

To obtain y_1, we substitute the series (i) in the ODE. This will lead us to

$$y_1 = J_0(x)$$

exactly along the lines of the solution to (1.41). To obtain y_2, we can either substitute (ii) in the ODE and obtain d_n's or use the Wronskian method. Here we shall exhibit the first of the alternatives. We note that

$$y_0 = \ln z \times y_1$$

satisfies

$$y_0' = \ln z \times y_1' + \frac{1}{z} y_1$$

$$y_0'' = \ln z \times y_1'' + \frac{2}{z} y_1' - \frac{1}{z^2} y_1$$

and thus,

$$y_0'' + \frac{1}{z} y_0' + y_0 = \ln z \times \left(y_1'' + \frac{1}{z} y_1' + y_1 \right) + \frac{2}{z} y_1' - \frac{1}{z^2} y_1 + \frac{1}{z^2} y_1 = \frac{2}{z} y_1'.$$

Consequently, $\tilde{y}_2$ satisfies

$$\tilde{y}_2 + \frac{1}{z} y_2' + \tilde{y}_2 = -\frac{2}{z} y_1' = -\frac{2}{z} J_0'(z)$$

$$= -\sum_{m=1}^{\infty} \frac{(-)^m}{m!(m-1)} \left(\frac{z}{2}\right)^{2(m-1)}.$$

We now substitute $\tilde{y}_2 = \sum_{n=1}^{\infty} d_n z^n$ in the above to obtain

$$\sum_{n=2}^{\infty} n(n-1) d_n z^{n-2} + \sum_{n=1}^{\infty} n d_n z^{n-2} + \sum_{n=1}^{\infty} d_n z^n = -\sum_{m=1}^{\infty} \frac{(-)^m}{m!(m-1)!} \left(\frac{z}{2}\right)^{2(m-1)}.$$

A comparison of the coefficients of the z^{-1} term on both sides leads to $d_1 = 0$. Comparison of coefficients of z^0,z^1 and z^2 lead to

$$4d_2 = 1, \quad 9d_3 = 0, \quad 14d_4 + d_2 = -\frac{1}{8}.$$

These imply $d_2 = 1/4$, $d_3 = 0$, $d_4 = -3/128$. Thus, the second linearly independent solution can be taken as

$$y_2(z) = \ln z \times y_1(z) + \frac{1}{4}z^2 - \frac{3}{128}z^4 + O(z^6).$$

(A suggested exercise: Study the behaviour of the above ODE and the two linearly independent solutions under $z \to -z$.)

1.6.3 An alternate method for the general solution in the vicinity of a regular singular point

Here, we shall discuss a different approach to the construction of the general solution in the vicinity of a regular singular point. As we have seen in the last section, the series solution method yields the general solution when the two roots of the indicial equation do not differ by an integer. The alternate method we are about to present naturally addresses itself just to the more 'difficult' case of roots differing by an integer. This method is often simpler to execute compared to the Wronskian method, and *simultaneously* yields two linearly independent solutions. It is also less cumbersome and more suitable when we want to evaluate the leading coefficients of the series for the second linearly independent solution *explicitly*.

We shall illustrate the method with a simple example. We shall again consider the Bessel equation. We know from our earlier discussion that the only cases where the series method does not yield the general solution are (i) $\nu = 0$; and (ii) $\nu = n$, a positive integer. We shall use precisely these cases to illustrate this method.

The Bessel equation reads

$$z^2y'' + zy' + (z^2 - \nu^2)y = 0. \tag{1.53}$$

Suppose, we now attempt a series solution of the form

$$y(z) = z^r \sum_{n=0}^{\infty} c_n z^n, \tag{1.54}$$

(where r is as yet arbitrary). We will be lead to an infinite set of conditions on the infinite set of coefficients $(c_0, c_1, c_2, \ldots\ldots c_n, \ldots\ldots)$ as before. These conditions, of course, are obtained by equating the coefficients of the successively higher powers of z to zero in the left-hand side of (1.53). Of these, the first one involving c_0 (obtained by equating the coefficient of z^r to zero) is the indicial equation, which would imply $r^2 = \nu^2$. Now, suppose that in the series (1.54) we require that this set of infinite conditions, *except this first one*, are fulfilled. Then r will still remain arbitrary but this set of conditions will successively determine $(c_1, c_2, \ldots\ldots c_n, \ldots\ldots)$ in terms of c_0, r and ν. We shall now consider the series (1.54) so restricted by these relations, viz.,

$$y(z) = z^r \sum_{m=0}^{\infty} c_0 c'_m(r, \nu, m) z^m. \tag{1.55}$$

Here, we have re-expressed c_m's in (1.54) as

$$c_m \equiv c_0 c'_m(r, \nu, m)$$

with the $c'_m(r, \nu, m)$ as the *known* functions, which have been determined via the recurrence relations. If we were to set $r = \nu = n$ in (1.55), we would simply recover the series solution proportional to $J_n(z)$. Evidently, for an arbitrary r, (1.55) will *not* satisfy the Bessel equation since the first of the infinite set of conditions required for it, the indicial equation, has not been imposed yet. So, when (1.55) is substituted in the left-hand side of (1.53), we do *not* get zero for the result, but a result proportional to the z^r-term. We thus find that y of (1.55) satisfies

$$z^2 y'' + zy' + (z^2 - \nu^2)y = c_0 z^r (r^2 - \nu^2). \tag{1.56}$$

(We iterate in passing that the coefficient of this z^r-term is such as would lead to the indicial equation, were it equated to zero).

Starting with this observation, we shall now proceed to discuss separately the two cases (i) $\nu = 0$; and (ii) $\nu = n$, a positive integer.

Case (i): $\nu = 0$

The equation (1.56) becomes

$$z^2 y'' + zy' + z^2 y = c_0 z^r r^2. \tag{1.57}$$

Now, as mentioned earlier, y of equation (1.55) is *not* a solution of (1.53) (with $\nu = 0$), as evinced by the non-vanishing right-hand side of (1.57). But we would like to employ (1.57) to construct a solution to the Bessel's equation that is linearly independent of $J_0(z)$. To do this, we differentiate the equation (1.57) with respect to r, and then set $r = 0$. We note that, on account the factor r^2, the right-hand side always yields zero when this operation is carried out. As for the left-hand side, we note that y is being looked upon as a function of two independent variables, r and z and the partial differentiations with respect to these commute. We thus obtain

$$z^2 \left(\frac{\partial y}{\partial r}\bigg|_{r=0} \right)'' + z \left(\frac{\partial y}{\partial r}\bigg|_{r=0} \right)' + z^2 \left(\frac{\partial y}{\partial r}\bigg|_{r=0} \right) = 0. \tag{1.58}$$

Evidently, the above equation implies that the quantity $\frac{\partial y}{\partial r}|_{r=0}$ is a solution of the Bessel equation with $\nu = 0$. To convince oneself that this solution we have found is in fact *linearly independent* of $J_0(z)$, we note the form of $\frac{\partial y}{\partial r}|_{r=0}$. From (1.55) we find

$$y_2 \equiv \left.\frac{\partial y}{\partial r}\right|_{r=0} = (\text{const}) \ln z J_0(z) + \sum_{m=0}^{\infty} c_0 \frac{\partial}{\partial r} c'_m(r, \nu = 0, m)\bigg|_{r=0} z^m.$$

The first term has come from the differentiation of z^r with r, while the second term is a power series. (Recall that if $r = \nu = 0$ is substituted in (1.55), y becomes proportional to the first solution $J_0(z)$). The presence of the logarithmic term immediately makes it evident that y_2 and J_0 are linearly independent of each other, because otherwise they would have to be proportional (see also Exercise 1.28).

Case (ii): $\nu = n$, a positive integer

We shall next develop a method similar to the above one for the case when ν is a positive integer. Suppose, again, we follow the procedure that lead to equation (1.56). Now, suppose we were to differentiate (1.56) with respect to r and set $r = -\nu$, we shall find that the right-hand side of (1.56) does contribute. This difference arises from the fact that while for the case of $\nu = 0$, the right-hand side of (1.57) was *quadratic* in $(r + \nu) \equiv r$, for the case of $\nu \neq 0$, the right-hand side of (1.56) is only linear in $(r + \nu)$. To remedy this situation, consider the replacement $c_0 \to b(r + \nu)$ throughout, with b a finite constant. Then, equation (1.56) becomes

$$z^2 y'' + zy' + (z^2 - \nu^2)y = bz^r(r - \nu)(r + \nu)^2. \tag{1.59}$$

We now note that the right-hand side is *quadratic* in $(r + \nu)^2$ analogous to (1.57). We can now follow the procedure of differentiation of (1.59) with respect to r and then set $r = -\nu = -n$, to find

$$z^2 \left(\left.\frac{\partial y}{\partial r}\right|_{r=-n}\right)'' + z\left(\left.\frac{\partial y}{\partial r}\right|_{r=-n}\right)' + (z^2 - n^2)\left(\left.\frac{\partial y}{\partial r}\right|_{r=-n}\right) = 0. \tag{1.60}$$

We thus arrive at the following result: Suppose we start from the series

$$y(z) = \sum_{m=0}^{\infty} b(r + \nu)c'_m(r, \nu, n)z^{m+r} \tag{1.61}$$

and compute $y_2 \equiv \left.\frac{\partial y}{\partial r}\right|_{r=-n}$, then y_2 will satisfy the Bessel equation with $\nu = n$. Further, we can compute y_2 as

$$y_2(z) = \left.\frac{\partial y}{\partial r}\right|_{r=-n} = \frac{\partial}{\partial r}\left[\sum_{m=0}^{\infty} b(r + \nu)c'_m(r, \nu, m)z^{m+r}\right]\Bigg|_{r=-n} \tag{1.62}$$

Some of the subtleties associated with (1.62) are best illustrated with the help of the following example.

Solved problem 1.13 Using the method outlined above, obtain a set of two linearly independent solutions for the Bessel equation for the two cases: (i) $\nu = 0$, (ii) $\nu = n$, a positive integer. Evaluate the logarithmic term and the powers up to $O(z^{-\nu+4})$.

Solution:

Case (i): $\nu = 0$

We substitute the series

$$y(z) = z^r \sum_{n=0}^{\infty} c_n z^n \tag{1.63}$$

in the Bessel equation with $\nu = 0$, viz.,

$$z^2 y'' + zy' + z^2 y = 0 \tag{1.64}$$

and equate the coefficients of various powers of z to obtain (see Section 1.6.1):

$$c_0 r^2 = 0$$

$$c_1 = c_3 = \ldots\ldots = 0$$

$$c_{2m} = c_{2m-2}\left[\frac{-1}{(2m+r)^2}\right]$$

Now, as elaborated while evolving the method, we shall impose all these relations on the coefficients except the first one, viz., $c_0 r^2 = 0$. This leads us to the expression for c_{2m} as

$$c_{2m} = c_0 \frac{(-)^m}{(2+r)^2(4+r)^2 \ldots\ldots\ldots (2m+r)^2} \equiv c_0 c'_{2m}(r, m).$$

We thus obtain for the series y

$$y(z) = z^r \sum_{m=0}^{\infty} c_0 c'_{2m}(r, m) z^{2m}. \tag{1.65}$$

Now, as shown earlier,

$$y_2 \equiv \left.\frac{\partial y}{\partial r}\right|_{r=0} = (\text{const}) \ln z J_0(z) + \sum_{m=0}^{\infty} c_0 \left.\frac{\partial}{\partial r} c'_{2m}(r, \nu = 0, m)\right|_{r=0} z^{2m}$$

constitutes a second linearly independent solution. From our knowledge of $c'_{2m}(r, m)$, we can now evaluate the terms up to $O(z^4)$. We find

$$y_2(z) = c_0 \ln z J_0(z) + c_0 \left(\frac{1}{4} z^2 - \frac{1}{64}\left(1 + \frac{1}{2}\right) z^4 + \cdots \right).$$

One can compare this solution with that obtained for this case by the Wronskian method in Section 1.6.2.

Case (ii): $\nu = n$, a positive integer

Again, we substitute the series

$$y(z) = z^r \sum_{n=0}^{\infty} c_n z^n \tag{1.66}$$

into the Bessel equation

$$z^2 y'' + z y' + (z^2 - \nu^2) y = 0 \tag{1.67}$$

to obtain the set of recurrence relations (see Section 1.6.1):

$$c_0(r^2 - \nu^2) = 0$$

$$c_1 = c_3 = \ldots\ldots = 0$$

$$c_{2m} = c_{2m-2} \frac{-1}{(2m + r)^2 - n^2}$$

As before, we do *not* impose the indicial equation $c_0(r^2 - \nu^2) = 0$. The remaining equations lead to

$$c_{2m} = c_0 \frac{(-)^m}{[(2 + r)^2 - n^2]\,[(4 + r)^2 - n^2] \ldots\ldots [(2m + r)^2 - n^2]} \equiv c_0 c'_{2m}(r, m).$$

Having obtained $c'_{2m}(r, m)$, we now replace c_0 by $b(r + n)$. This implies, $c_{2m} \to b(r + n) c'_{2m}(r, m)$ in (1.66). We thus write (1.66) as

$$y(z) = \sum_{m=0}^{\infty} b(r + n) c'_{2m}(r, m) z^{2m+r}. \tag{1.68}$$

At this point we draw attention to the presence of the m factors occurring in the denominator of c'_{2m} each of the form $[(2s + r)^2 - n^2]$. We note that if we set $r = -n$, none of these factors can vanish except the one with $s = n$. Since this factor is present (exactly once) in the denominator of every c'_{2m} with $m \geq n$, each of these blows up at $r = -n$. Thus, recalling the presence of the multiplicative factor $(r + n)$ in the above equation, we note that

$$\lim_{r\to -n} (r+n)c'_{2m} = 0; \quad m < n$$
$$= \text{finite} \quad m \geq n. \tag{1.69}$$

Thus, if we were just to take the limit $r \to -n$ in (1.68), we would find that the first n terms of the series vanish and the series begins as z^n. This solution can easily be verified to be proportional to $J_n(x)$.

Keeping the above in mind, we differentiate the right-hand side of (1.68) with respect to r and set $r = -n$ to find

$$y_2(z) \equiv \left.\frac{\partial y}{\partial r}\right|_{r=-n} = \ln z \sum_{m=0}^{\infty} b[(r+n)c'_{2m}(r,\nu,m)]\Big|_{r=-n} z^{2m-n} \tag{1.70}$$

$$+\sum_{m=0}^{\infty} b\frac{\partial}{\partial r}[(r+n)c'_{2m}(r,\nu,m)]\Big|_{r=-n} z^{2m-n} \tag{1.71}$$

$$= \ln z \times (\text{const})\, J_n(z) + (\text{a series starting as } z^{-n}). \tag{1.72}$$

We note that the form of $y_2(z)$ enables us to see that it cannot be proportional to $J_n(z)$ which behaves as z^n near $z = 0$. Thus, y_2 constitutes a solution linearly independent of $J_n(z)$. Knowing the form of c'_{2m}, we can evaluate both the constant in the first term and the first three terms in the series. For $n > 2$, they are

$$y_2(z) = (\text{const}) \ln z J_n(z) + z^{-n}\left(1 - \frac{z^2}{4(1-n)} + \frac{z^4}{32(1-n)(2-n)} + O(z^6)\right). \tag{1.73}$$

1.7 Solutions to the ODE around an Irregular Singular Point

We close the discussion on second order homogeneous ODEs that occur in a physical context by making some remarks about the solutions to such an equations about an irregular singular point. We shall still restrict our attention to the case when both p and q have an *isolated* singularity at $z = z_0$. Here we consider two cases:

(a) p has a pole of order higher than one.
(b) p has at the most a pole of order one, and q has a pole of order more than two.

In case (a), we have already shown (see Theorem 1.1) that $I(z)$ has an essential singularity at $z = z_0$. We recall that the Wronskian of two solutions, assumed to exist, is (modulo a constant)

$$W(z) = I^2(z)$$

and thus necessarily has an essential singularity at $z = z_0$. Thus, it follows that *at least one of the solutions* will have an essential singularity at $z = z_0$. Moreover, on account of the fact that that $I(z)$ has an essential singularity,

$$y(z) = I(z)f(z)$$

itself has one, barring exceptional cases in which it cancels between two factors (for an example and further discussion, see below).

In case (b), we find that $I(z)$ will *not* have an essential singularity. Nonetheless, consider the equation

$$f'' + Qf = 0.$$

Now, Q will have a pole of order three or higher. In this case it is not difficult to verify that there *cannot* be a solution of the form

$$f = (z - z_0)^r \sum_0^\infty c_n (z - z_0)^n.$$

(See Exercise 1.44)

Solved problem 1.14 Consider the equation $y'' + \frac{1}{z^2}y' - \frac{1}{z^3}y = 0$. Show that

(a) the ODE has an irregular singular point at $z = 0$.
(b) $y_1 = z$ is a solution.
(c) The Wronskian $W(z)$ is given by $W(z) = C\exp(1/z)$.
(d) The second linearly independent solution has an essential singularity at $z = 0$.

Solution:

(a) Here, p and q have poles of order 2 and 3 respectively at $z = 0$. Thus, $z = 0$ is an irregular singular point.
(b) By inspection.
(c) The Wronskian is given by

$$W(z) = C\exp\left\{-\int^z dz' \frac{1}{z'^2}\right\} = C\exp(1/z).$$

(d) We can write the second linearly independent solution by the Wronskian method:

$$y_2(z) = y_1(z)\int^z \frac{W(z')}{y_1^2(z')} dz'$$

$$= z \int^{z} \frac{\exp(1/z')}{z'^2} dz'$$

$$= -\sum_{0}^{\infty} \frac{1}{z^n (n+1)!} \tag{1.74}$$

We note that the principal part of $y_2(z)$ has an infinite number of terms implying the existence of an essential singularity at $z = 0$.

Solved problem 1.15 Consider again the same ODE $y'' + \frac{1}{z^2}y' - \frac{1}{z^3}y = 0$. Show that if one substitutes a series

$$y = \sum_{0}^{\infty} c_n z^{n+r}$$

in the ODE, there is only *one* solution for the index r.

Solution: We substitute the above series in the ODE and compare the coefficient of the lowest power of z, viz., z^{r-3}. We find that unlike the case of the regular singular point, there is no contribution from the y'' term and the indicial equation is *linear* in r and reads $r = 1$.

Solved problem 1.16 Consider the ODE $y'' + \frac{1}{z^2}y' - \frac{1}{z^4}y = 0$. Show that one cannot have a solution of the form of a series

$$y = \sum_{0}^{\infty} c_n z^{n+r}$$

for the ODE for any r.

Solution: A substitution of the series in the ODE and the comparison of the lowest power z^{r-4} leads us to the equation, $c_0 = 0$. By assumption, $c_0 \neq 0$. Thus the series for any r cannot satisfy the ODE.

To summarise, in the case of the solution of the ODEs under consideration about an irregular singular point, at least one of the solutions, and hence the general solution, has an essential singularity.

1.7.1 Behaviour of a function near an essential singularity

To understand the relevance or otherwise of a function with an essential singularity from the physicist's perspective, we need to note some properties of a function near its essential singularity. It can be shown that[9] in the vicinity of an essential singularity, the behaviour of a function is extremely irregular and wild. This is expressed by Picard's theorem:

[9]See e.g., *Complex Analysis and Applications* by Churchill and Brown, 4th Edition, Sections 56 and 107.

In *each neighbourhood* of an essential singular point, the function assumes *every finite value*, with one possible exception, *an infinite number* of times.

We do not expect physical quantities to behave in such an irregular fashion. Consequently, we cannot expect Physics to involve equations which have irregular singular points in the *finite* complex plane. The table in Section 1.4.2 confirms this. [On the other hand, the table shows that there are equations in it which have an irregular singular point at *infinity*. But this does not invalidate the above argument (Prove!)].

Exercises

On §1.2

1.1 Show that a solution of

$$(1-x^2)y''(x) - 2xy'(x) + \left[l(l+1) - \frac{m^2}{1-x^2}\right]y(x) = 0$$

can be written as

$$y(x) = (1-x^2)^{m/2}\frac{d^m}{dx^m}Y(x),$$

where $Y(x)$ is a solution of

$$(1-x^2)Y''(x) - 2xY'(x) + l(l+1)Y(x) = 0.$$

On §1.3

1.2 Prove the properties of the Wronskian $W[y_1, y_2]$ of two functions y_1 and y_2 listed in Section 1.3.3.

1.3 Show that the Wronskian of $\tilde{y}_1 = 5\sin^2 z - \cos^2 z$ and $\tilde{y}_2 = \sin^2 z - 3\cos^2 z$ is equal to (-14) times that of $y_1 = \sin^2 z$ $y_2 = \cos^2 z$. Show further that the latter equals the Wronskian of $\left[-\frac{1}{2}\cos 2z\right]$ and 1. Evaluate it.

1.4 Show that the Wronskian of two linearly independent functions $y_1(z)$ and $y_2(z)$ is analytic at $z = z_0$ if both $y_1(z)$ and $y_2(z)$ are analytic there. Construct an example to show that the converse need not be true.

1.5 Show that the Wronskian of n entire functions either vanishes everywhere or has only isolated zeros.

1.6 Evaluate the Wronskian of the n functions $\left\{\frac{z^n}{n!}; (n = 1, 2, \ldots, N)\right\}$, and show that the set of functions are linearly independent.

1.7 Using the Wronskian method, show that the three functions $f_m(z) = \sin mz$ $(m = 1, 2, 3)$ are linearly independent.

1.8 Find the Wronskian for the differential equation up to an arbitrary constant:

$$(x-1)(x-2)y'' + y' - \frac{y}{x} = 0.$$

One solution to the above differential equation is obtained as $y_1 = -x$. Obtain a second linearly independent solution in a closed form.

1.9 Show that the most general ODE of the form

$$y'' + p(z)y' + q(z)y = 0$$

for which $y = z$ is a solution is of the form

$$y'' + p(z)[y' - \frac{1}{z}y] = 0$$

In this case, the second linearly independent solution is given by

$$y_2(z) = z\int^z \frac{dz'}{z'^2}\exp\left[-\int^{z'} p(z'')dz''\right].$$

1.10 Show that the most general ODE of the form

$$y'' + p(z)y' + q(z)y = 0$$

for which $y_1 = \text{const}$ is one of the solutions is of the form

$$y'' + p(z)y' = 0.$$

Show that this can be directly integrated to yield (up to a multiplication constant)

$$y' = \exp\left[-\int^z p(z')dz'\right].$$

This yields

$$y = \int^z dz' \exp\left[-\int^{z'} p(z'')dz''\right]$$

up to a constant. Show that the above is also the second linearly independent solution y_2 obtained by the Wronskian method.

On §1.4

1.11 Verify the singular points and the nature for the differential equations listed in Table 1.1 of Section 1.4.2.

1.12 Consider the ODE: $y'' + p(z)y' + q(z)y = 0$ with p having a pole of an order two or higher at $z = 0$. Evaluate $I(z)$ of (1.26) and show that it has an essential singularity at $z = 0$. Consider now the same ODE with p having no singularity worse than a simple pole and q having a pole of the order three or higher at $z = 0$. Show, in this case, that $I(z)$ has no essential singularity at z. Show further that $Q(z)$ of (1.28) has the same worst singularity as $q(z)$. Show in this case that (1.25) has no solution analytic at $z = 0$, and if (1.25) has a solution with an isolated singularity at $z = 0$, then it must be an essential singularity.

1.13 Using the form (1.26), show that if $p(z)$ has a double pole at $z = 0$, then $y(z)$ *generally* has an essential singularity at $z = 0$. When would this result fail? Construct a counter-example.

On §1.5

1.14 Show that the differential equation $y'' + p(z)y' + q(z)y = 0$ must have at least one singular point in the extended complex plane. Recall that the extended complex plane has the point at infinity augmented to the complex plane.
[Hint: Use the result that an entire function that is bounded is a constant (Liouville's Theorem)].

1.15 We have cast $y'' + p(z)y' + q(z)y = 0$ in the form $f'' + Qf = 0$ by a change of variables in (II).

(a) Show that the change of variables can be written as

$$y = \text{const} \times \sqrt{W(z)}\, f.$$

(b) Show that the two linearly independent solutions of $f'' + Qf = 0$ have a constant Wronskian.

(c) Verify that the results in (a) and (b) are compatible with the Wronskian of any two linearly independent solutions of the ODE being $W[z]$.

1.16 An ODE $y'' + py' + qy = 0$ has only one singular point at infinity. Show then that there are two linearly independent entire solutions y_1 and y_2.

1.17 Consider the pair of equations

$$y'' + py' + qy = 0 \tag{I}$$

and

$$f'' + Qf = 0 \tag{II}$$

as used in Section 1.5. Show that if $z = 0$ is an ordinary point of (I), then it is also an ordinary point of (II). Find the conditions under which a regular singular point of (I) is (i) a regular singular point of (II) and (ii) an ordinary point of (II). Construct an example of each case.

1.18 The series on the right-hand side of (1.30) of the upper bounds on $|f_n|$ seems to converge absolutely in the entire complex plane, being an exponential series, whereas the series for $f(z)$ may not converge outside a limited domain as the differential equation may have singular points. How do you reconcile?

1.19 You are given that $x = 0$ is an ordinary point of the differential equation: $y'' + xy' + q(x)y = 0$. Obtain the necessary and sufficient conditions under which the ODE will have two linearly independent solutions, one even and one odd.

1.20 Calculate the Wronskian for the two linearly independent solutions of the Hermite differential equation

$$y'' - 2xy' + 2ny = 0.$$

Use this *form* of the Wronskian (and without any explicit solution method) to argue that this equation cannot have more than one (linearly independent) solution that is a polynomial.

On §1.6

1.21 Consider the differential equation $4x^2y'' + y = 0$. Find the nature of the point at infinity. Write down the form of two linearly independent solutions to the equation about $x = \infty$. Obtain one of the solutions completely.

1.22 Consider the equation,

$$z(z-1)y'' + [(a+b+1)z - c]y' + aby = 0 \tag{1.75}$$

called the Hypergeometric equation. Show that the equation has three regular singular points at $z = 0, 1, \infty$. Show further that when $c \neq$ integer, the equation has two linearly independent solutions given by

$$\begin{aligned} y_1(z) &= F(a, b|c|z), \\ y_2(z) &= z^{1-c}F(a+1-c, b+1-c|2-c|z)), \end{aligned} \tag{1.76}$$

where, $F(a, b|c|z)$ is been defined by the series

$$F(a, b|c|z) \equiv \sum_0^\infty \frac{(a+n-1)!(b+n-1)!(c-1)!}{(a-1)!(b-1)!(c+n-1)!(n)!} z^n. \tag{1.77}$$

Show that the above series converges in the domain $|z| < 1$.

1.23 (a) Show that both the roots of the indicial equation are real if both a_0 and b_0 are real and $b_0 \leq 0$. (b) Verify that ALL the equations in the Table 1.1 in the Section 1.4.2 (whose solutions can directly occur in physical observables) have real roots to indicial equations about their regular singular points.

1.24 Consider the ODE

$$z(1-z)y''(z) + (c-z)y'(z) + 1/4y(z) = 0.$$

(a) Find the nature of the point at infinity.
(b) Find the regular singular points of the ODE.
(c) Write down the form of the general solution about $z = 0$ when (i) $c = 1/2$, (ii) $c = 1$, (iii) $c = 2$.
(d) Obtain the general solution explicitly for the case $c = 1$ and $c = 2$.

1.25 Consider the Legendre differential equation

$$(1-z^2)y'' - 2zy' + l(l+1)y = 0.$$

(a) Show that the above equation has a polynomial solution whenever l is a non-negative integer.
(b) Study the nature of the points $z = \pm 1$. Write down the indicial equation about $z = 1$ and show that at least one of the solutions diverges at $z = 1$.
(c) Now show that $z = 0$ is an ordinary point of the differential equation. Let y_1 and y_2 be two solutions analytic at $z = 0$. Show that at least one of these series must diverge at $z = \pm 1$.

1.26 Consider the Laguerre equation

$$zy'' + (1-z)y' + ny = 0.$$

(a) Find the nature of the point $z = 0$. Write down the indicial equation.
(b) Show that for certain values of n, *one* of the solutions is a *polynomial*.
(c) Can there be *two* polynomial solutions for any given n? [Hint: Consider the Wronskian].

1.27 Let $z = 0$ be a regular singular point of the ODE

$$y'' + p(z)y' + q(z)y = 0$$

with the indicial equation $r^2 = 0$. Show that this implies that in the vicinity of $z = 0$, p and q have the form

$$p(z) = \frac{1}{z} + \text{a function analytic at } z = 0.$$

$$q(z) = \frac{\alpha}{z} + \text{a function analytic at } z = 0.$$

Show further that, in this case, the two linearly independent solutions about $z = 0$ have the form

$$y_1 = \sum_{n=0}^{\infty} c_n z^n,$$

$$y_2 = y_1 \ln z + \sum_{n=1}^{\infty} d_n z^n.$$

In particular, note that the coefficient of the $y_1 \ln z$ term is *necessarily nonzero* in this case and the series in y_2 starts with z^1 (and not z^0).

1.28 In Section 1.6.3, we solved the Bessel equation for $\nu = 0$ and obtained the form of the two solutions. Show explicitly that they are linearly independent. Consideration of the behaviour near $z = 0$ will be helpful.

1.29 Consider the two differential equations

(i) $y'' + y' + y = 0,$

(ii) $y'' + y' + \frac{1}{4z^2} y = 0.$

Discuss the analytic properties of the solutions to both in the neighbourhood of $z = 0$. Show, in particular, that the Wronskian of two linearly independent solutions of both the differential equations is identical (up to a constant) and is analytic at $z = 0$ in both cases, irrespective of different analytic properties of the solutions in the two cases.

1.30 For the ODE discussed in this chapter, let $p(z)$ have a pole of order 1 at $z = 0$ with a residue a_0 and let $q(z)$ be analytic at $z = 0$. Show that one solution to the equation is analytic and the other linearly independent solution has a branch-cut there provided $a_0 \neq$ a negative integer.

1.31 Suppose that the ODE: $y'' + p(x)y' + q(x)y = 0$ is even under $x \to -x$ (i.e., it does not change the form).

(a) Show that if $y(x)$ is a solution then so is $y(-x)$.

(b) Suppose a solution of $y(x)$ is neither even nor odd under $x \to -x$, then $\frac{1}{2}(y(x) + y(-x))$ and $\frac{1}{2}(y(x) - y(-x))$ are the two linearly independent solutions.

(c) Suppose $x = 0$ is a regular singular point and a series solution $y = \sum_{n=0}^{\infty} c_n x^n$ is attempted, then show that any recurrence relation contains either only even coefficients or only odd coefficients. Hence show that if the two roots of the indicial equation r_1 and r_2 do not differ by an integer, there will be two linearly independent solutions of the form

$$y_1 = x^{r_1} \sum_{n=0}^{\infty} c_{2n} x^{2n}$$

and

$$y_2 = x^{r_2} \sum c'_{2n} x^{2n}.$$

(d) Suppose now that $x = 0$ is an ordinary point of the ODE. Show that two linearly independent solutions can be chosen to be even and odd each.

(e) Verify the above statements for the Hermite differential equation and the Bessel differential equation.

1.32 (a) Consider the ODE $f'' + Qf = 0$. Let two of its linearly independent solutions be f_1 and zf_1. Show that f_1 must a constant and $Q \equiv 0$.

(b) Find the general condition on p and q on which the ODE will have two linearly independent solutions whose ratio is z.

1.33 Consider the Bessel equation with $\nu = \frac{1}{2}$. $J_{1/2}(z)$ is one solution of the equation. Explain why the other linearly independent solution $J_{-1/2}(z)$ does not contain a term proportional to $\ln z J_{1/2}(z)$. Do this by trying a solution of the form

$$y_2 = c \ln z J_{1/2}(z) + \sum_{n=0}^{\infty} a_n z^{n-\frac{1}{2}}$$

and obtaining $c = 0$.

1.34 Consider the Bessel equation with $\nu = 0$. $J_0(z)$ is one solution. Obtain the first few terms of the second linearly independent solution by the Wronskian method and compare with the result in Section 1.6.3.

1.35 Show that the most general solution to the equation

$$x^2 y'' + x(1 - 2\alpha)y' + [\beta^2 \gamma^2 x^{2\gamma} + (\alpha^2 - n^2\gamma^2)]y = 0$$

is

$$y = x^\alpha [A J_n(\beta x^\gamma) + B Y_n(\beta x^\gamma)],$$

where A and B are arbitrary constants and J_n and Y_n are two linearly independent solutions of the Bessel equation. (Hint: Use an appropriate change of variables).

1.36 Obtain two linearly independent solutions of the following differential equation up to $O(x^4)$:

$$x^2 y'' + 2xy' - (2 + x^2)y = 0.$$

1.37 Construct a second order ordinary differential equation for which $z = 0$ is a regular singular point and has roots to the indicial equation $r_1 = 2$ and $r_2 = 1$. Is this equation unique? State without proof the form of two linearly independent solutions to this equation.

1.38 A second order ODE $y'' + py' + qy = 0$ is invariant under $z \to -z$ and it has $z = 0$ as a regular singular point. Further, the roots of the indicial equation are both integral: one even and one odd. Show then that the 'second' linearly independent solution does not have the $\ln z \times y_1$ term in it. Construct an example.

1.39 Write down the most general second order ODE:

$$y'' + p(z)y' + q(z)y = 0$$

that behaves homogeneously under $z \to \lambda z$ (λ = constant). (Hint: $p(z)$ and $q(z)$ must have the form $p(z) = \frac{\alpha}{z}, q(z) = \frac{\beta}{z^2}$). Write down the roots of the indicial equation for such an ODE. Write down the conditions under which (i) $r_1 = r_2$. (ii) $r_1 - r_2 = N =$ an integer. Show by direct substitution that if $r_1 \neq r_2$, z^{r_1} and z^{r_2} can be taken to be the two linearly independent solutions; and that this includes the case when $r_1 - r_2$ is a nonzero integer. Show further that if $r_1 = r_2 = r$, the two linearly independent solutions are z^r and $z^r \ln z$. Verify by the Wronskian method the above results. Study the behaviour of the above solutions under $z \to \lambda z$. Can you argue out these results from the fact that the ODE behaves homogeneously under $z \to \lambda z$?

1.40 Complete the derivation of (1.73).

1.41 Consider the ODE

$$z^2 y'' - 5zy' + 8y = 0.$$

Find the nature of point at $z = 0$. Find the roots of the indicial equation and show that they differ by an integer. Show however that the second solution does not have a logarithmic term. Show further that both the linearly independent solutions are analytic at $z = 0$. This, thus is an exception to the general 'rule' that at least one solution in the vicinity of a regular singular point is not analytic there.

1.42 Construct an example of an ODE with both solutions analytic at a regular singular point. (Hint: The two indices r_1 and r_2 must be +ve integers and the logarithmic term in the second solution must be missing.)

On §1.7

1.43 Consider the ODE

$$z^4 y'' + 2z^3 y' + y = 0.$$

Find the nature of the point at $z = 0$. Show that one of the solutions to the ODE is $y = \exp(1/z)$. Study the behaviour of y near the singularity $z = 0$ by writing $z = \epsilon \exp(i\phi)$. Show that on the small circle $|z| = \epsilon$, $|y|$ and $\arg y$ both vary very wildly. Can one generally expect such a function to appear in the solution for y if y represents a *physical quantity*? Comment on the connection between this and such considerations and the fact that the singular points (within the finite complex plane) of all the ODES that commonly occur in physics are only of the *regular singular* variety?

1.44 Consider the ODE

$$f'' + Qf = 0$$

with Q having a pole of order three at $z = 0$. Attempt a solution of the form

$$y = \sum_{0}^{\infty} c_n z^{n+r}.$$

Show that there is *no* solution possible of the above form.

References

[1] Whittaker, E. T. and Watson, G. N., *A Course of Modern Analysis: An Introduction to the General Theory of Infinite Processes and of Analytic Functions, with an Account of the Principal Transcendental Functions*, AMS Press, New York, 1979.

[2] Courant, Richard and Hilbert, David, *Methods of Mathematical Physics*, Interscience Publishers, New York, 1989.

[3] Morse, Philip McCord and Feshbach, Herman, *Methods of Theoretical Physics*. Vol. I-II, McGraw-Hill, Boston, Mass, 1999.

[4] Sommerfield, A., *Partial Differential Equations in Physics*, Academic Press, Inc., New York, 1949.

2 Gamma (Factorial) and Beta Functions

In this chapter we shall study the generalisation of the gamma (or the factorial) function to the one with a complex argument. To begin with, let us recall the familiar factorial function $n!$ for the nonnegative integer arguments:

$$n! = 1 \cdot 2 \cdots (n-1) \cdot n$$
$$0! \equiv 1$$

We widely encounter a generalisation of the above, viz. the factorial function $z!$ (or the gamma function $\Gamma(z+1) \equiv z!$) with noninteger or even complex arguments both in mathematics in general, as well as in several physics applications. The manner in which the gamma function frequently enters is via the evaluation of one of the definite integrals of the type

(a) $\int_0^\infty e^{-t} t^\alpha dt$ $\qquad$ $(\text{Re}\,\alpha > -1)$

(b) $\int_0^1 dt\, t^\alpha (1-t)^\beta$ $\qquad$ $(\text{Re}\,\alpha > -1, \text{Re}\,\beta > -1)$

and also the several related integrals such as

(c) $\int_0^\infty e^{-t^2} t^m dt$ $\qquad$ $(\text{Re}\,m > -1)$

(d) $\int_0^{\pi/2} d\theta \sin^m\theta \cos^n\theta$ $\qquad$ $(\text{Re}\,m > -1, \text{Re}\,n > -1)$

(e) $\int_1^\infty \frac{u^\alpha}{(1+u)^\beta} du$ $\qquad$ $(\text{Re}\,(\alpha - \beta) < -1)$

that one encounters in various contexts in mathematics and physics. They enter also into the series definition of Bessel functions. Certain advanced areas of physics such as quantum field theory (with applications to high energy physics, condensed matter physics and statistical physics) use analytic continuation of integrals to complex dimensions, and their evaluation involves the gamma function.

2.1 Definition

We shall first present a heuristic approach to developing a definition of the factorial function $z!$ for a complex argument z. In particular, for $z = n$, a positive integer, this definition must reduce to $n!$. If we require that $z!$ be analytic (in as large a domain as possible) and to reduce to $n!$ when $z = n$, it is not sufficient to uniquely determine $z!$ (See Exercises 2.26 and 2.27). To further define such a function, one needs to adopt additional guidelines. Suppose we require that we preserve the complex analogue of $n! = n(n-1)!$ i.e., we should require $z!$ to satisfy $z! = z(z-1)!$ for all z. For a given z, this enables one to relate to each other the factorial function for all arguments of the form $(z+m)$, where m is an integer (assuming $(z+m)!$ exists). We can now write

$$z! = \frac{1}{z+1}(z+1)! = \frac{1}{(z+1)(z+2)}(z+2)! \tag{2.1}$$

$$= \frac{1}{(z+1)(z+2)\cdots(z+n)}(z+n)! \tag{2.2}$$

for an arbitrary n. (2.2) is true in particular as $n \to \infty$. (2.2) would give us a way to define $z!$ if we add information about the behaviour of $(z+n)!$ as $n \to \infty$. We add this information by looking at the behaviour of $(z+n)!$ for $z = m$ a positive integer. We note

$$(n+m)! = n!(n+1)(n+2)\cdots(n+m)$$

$$= n!\,n^m\left(\frac{n+1}{n}\right)\cdots\left(\frac{n+m}{n}\right).$$

Thus,

$$\lim_{n\to\infty}\frac{(n+m)!}{n!\,n^m} = 1. \tag{2.3}$$

The above is true for a +ve integer m. We now adopt the above known result to an arbitrary value of z:

$$\lim_{n\to\infty}\frac{(n+z)!}{n!\,n^z} = 1. \tag{2.4}$$

This then enables us to write (2.2) as

$$z! = \lim_{n\to\infty}\left\{(n+z)!\prod_{r=1}^{n}\left(\frac{1}{z+r}\right)\right\}$$

$$= \lim_{n\to\infty}\left\{\frac{(n+z)!}{n!\,n^z}n!\,n^z\prod_{r=1}^{n}\left(\frac{1}{z+r}\right)\right\}$$

$$= \lim_{n\to\infty}\frac{n!\,n^z}{(z+1)\cdots(z+n)}, \tag{2.5}$$

where we have made use of (2.4)[10]. This, of course, is meaningful provided the above limit exists; this will be shown in Section 2.3.

[10]We have defined $n^z = \exp\{z \text{ Log } n\}$. Note that here, n is a positive integer and $\text{Log}\, n$ is the usual real (nonnegative) logarithm (the principal value).

Two different notations are commonly used in literature for this function. In addition to the factorial function $z!$ already introduced, we also use the *gamma* function defined below.

Definition: We define the Γ function by

$$\Gamma(z) \equiv (z-1)! \tag{2.6}$$

$$= \lim_{n\to\infty} \frac{n!\, n^{(z-1)}}{z\cdot(z+1)\cdots(z+n-1)} \tag{2.7}$$

for any z.

2.2 General Properties of $z!$ (and the Gamma Function)

Starting from the definition of $z!$ arrived at in the last section, viz.

$$z! = \lim_{n\to\infty} \frac{n!\, n^{z}}{(z+1)(z+2)\cdots(z+n)}, \tag{2.8}$$

we can deduce a number of properties.

1. Recurrence relation

$$z! = z(z-1)!$$

To see this, consider

$$\begin{aligned} z(z-1)! &= z \lim_{n\to\infty} \frac{n!\, n^{z-1}}{z(z+1)\cdots(z-1+n)} \\ &= \lim_{n\to\infty} \frac{n!\, n^{z}(1/n)(z+n)}{(z+1)(z+2)\cdots(z-1+n)(z+n)} \\ &= \lim_{n\to\infty} \frac{n!\, n^{z}}{(z+1)(z+2)\cdots(z+n)} \lim_{n\to\infty} \frac{z+n}{n} \\ &= z! \times 1 = z! \end{aligned}$$

We recall that we had, in fact, constructed the definition of $z!$ so as to have this property built into it.

2. Poles and Residues $z!$ has simple poles at $z = -m$ (m is a positive integer) with residue $(-)^{m-1}/(m-1)!$

To see this, consider

$$(z+m)z! = (z+m) \times \lim_{n\to\infty} \frac{n!\,n^z}{(z+1)(z+2)\cdots(z+m-1)(z+m)(z+m+1)\cdots(z+n)}$$

$$= \lim_{n\to\infty} \frac{n!\,n^z}{(z+1)(z+2)\cdots(z+m-1)(z+m+1)\cdots(z+n)}$$

Now at $z=-m$ the right-hand side becomes

$$= \lim_{n\to\infty} \frac{n!\,n^{-m}}{(-m+1)(-m+2)\cdots(-1)\cdot 1\cdot 2\cdot\ \cdots(n-m)}$$

$$= \frac{(-)^{m-1}}{(m-1)!} \lim_{n\to\infty} \frac{n!\,n^{-m}}{(n-m)!}$$

The last limit exists, and from (2.3), it is equal to one (set $n+m\to n$ in (2.3)). Hence, the right-hand side exists at $z=-m$. Thus,

$$\lim_{z\to -m} (z+m)z! = \frac{(-)^{m-1}}{(m-1)!}$$

Thus, $z!$ has a simple pole at $z=-m$, $(m=1,2,3,\dots)$ with a residue $(-)^{m-1}/(m-1)!$. Equivalently, $\Gamma(z)$ has a simple pole at $z=-m$, $(m=0,1,2,\dots)$ with residue $\frac{(-)^m}{m!}$

3. Duplication formula One of the remarkable properties of the factorial function is that it is possible to relate factorial of z to factorial of $2z$, i.e., the function with twice the argument. This, for obvious reasons, is called the duplication formula. It reads

$$(2z)! = \frac{z!\left(z-\frac{1}{2}\right)!\,2^{2z}}{\left(-\frac{1}{2}\right)!}.$$

To prove this result we shall write

$$z! = \lim_{n\to\infty} \frac{n!\,n^z}{(z+1)(z+2)\cdots(z+n)}$$

$$\left(z-\frac{1}{2}\right)! = \lim_{n\to\infty} \frac{n!\,n^{z-\frac{1}{2}}}{\left(z+\frac{1}{2}\right)\left(z+\frac{3}{2}\right)\cdots\left(z+n-\frac{1}{2}\right)}$$

$$\left(-\frac{1}{2}\right)! = \lim_{n\to\infty} \frac{n!\,n^{-\frac{1}{2}}}{\left(\frac{1}{2}\right)\left(\frac{3}{2}\right)\cdots\left(n-\frac{1}{2}\right)}$$

Now, we shall express $(2z)!$ with a somewhat modified appearance. In the definition of $(2z)!$, we shall use $(2n)$ as the parameter that tends to infinity. This is permissible since the limit exists irrespective of whether this parameter is even or odd. In other words, we shall write

$$(2z)! = \lim_{n\to\infty} \frac{(2n)!(2n)^{2z}}{(2z+1)(2z+2)\cdots(2z+2n)}.$$

Now we consider the quantity

$$\frac{2^{-2z}(2z)!\left(-\frac{1}{2}\right)!}{z!\left(z-\frac{1}{2}\right)!} = 2^{-2z} \lim_{n\to\infty} \frac{(2n)!}{n!} 2^{2z}$$

$$\times \frac{(z+1)(z+2)\cdots(z+n)\left(z+\frac{1}{2}\right)\cdots\left(z+n-\frac{1}{2}\right)}{(2z+1)\times\cdots\times(2z+2n)\times\frac{1}{2}\cdot\frac{3}{2}\cdots\left(n-\frac{1}{2}\right)} \tag{2.9}$$

Here we have used the *same* integer n for convenience in all the four limits[11] and have cancelled some factors. We write the above as

$$\lim_{n\to\infty} A_n \times B_n(z),$$

where

$$A_n = \frac{(2n)!}{n!\left(\frac{1}{2}\right)^n 1.3\cdots(2n-1)}$$

and

$$B_n(z) = \frac{(z+1)(z+2)\cdots(z+n)\left(z+\frac{1}{2}\right)\cdots\left(z+n-\frac{1}{2}\right)}{(2z+1)(2z+2)\cdots(2z+2n)}.$$

Evidently,

$$A_n = \frac{[2\cdot 4\cdot 6\cdot\ \cdots\ \cdot 2n]\times[1\cdot 3\cdot\ \cdots(2n-1)]}{[1\cdot 2\cdot 3\cdots n]\times\left(\frac{1}{2}\right)^n\times[1\cdot 3\cdot\cdots\cdot(2n-1)]} = 2^n\cdot 2^n = 2^{2n}.$$

$$B_n(z) = \left[\frac{(z+1)(z+2)\cdots(z+n)}{(2z+2)\cdots(2z+2n)}\right]\left[\frac{\left(z+\frac{1}{2}\right)\cdots\left(z+n-\frac{1}{2}\right)}{(2z+1)\cdots(2z+2n-1)}\right]$$

$$= \left(\frac{1}{2}\right)^n\left(\frac{1}{2}\right)^n = 2^{-2n}.$$

---

[11]We have employed a generalisation of $\lim_{n\to\infty} X_n Y_n = \lim_{n\to\infty} X_n \lim_{n\to\infty} Y_n$ provided both limits on the right-hand side exist.

Thus $A_n B_n = 1 = \lim_{n\to\infty} A_n B_n$. This shows

$$\frac{2^{-2z}(2z)!\left(-\frac{1}{2}\right)!}{z!\left(z-\frac{1}{2}\right)!} = 1.$$

In other words,

$$(2z)! = \frac{2^{2z}z!\left(z-\frac{1}{2}\right)!}{\left(-\frac{1}{2}\right)!}$$

We shall later show that $(-\frac{1}{2})! \equiv \Gamma(\frac{1}{2}) = \sqrt{\pi}$. This leads us to

$$(2z)! = 2^{2z}z!\left(z-\frac{1}{2}\right)!\pi^{-\frac{1}{2}}. \tag{2.10}$$

The duplication formula may alternately be expressed in terms of the Γ function. We can re-express (2.10) as

$$\Gamma(2z+1) = 2^{2z}\Gamma(z+1)\Gamma\left(z+\frac{1}{2}\right)\pi^{-\frac{1}{2}}.$$

Using $\Gamma(z) = (z-1)\Gamma(z-1)$, etc., this leads to

$$2z\Gamma(2z) = 2^{2z}z\Gamma(z)\Gamma\left(z+\frac{1}{2}\right)\pi^{-\frac{1}{2}},$$

i.e.,

$$\Gamma(2z) = 2^{2z-1}\Gamma(z)\Gamma\left(z+\frac{1}{2}\right)\pi^{-\frac{1}{2}}. \tag{2.11}$$

4. Relation between $z!\,(-z)!$ Another interesting property of $z!$ relates $z!$ to the factorial function with the negative argument $-z$. It reads

$$(z)!(-z)! = \frac{\pi z}{\sin \pi z}.$$

Proof: We note

$$z! = \lim_{n\to\infty}\frac{n!\,n^z}{(z+1)(z+2)\cdots(z+n)} = \lim_{n\to\infty}\frac{n^z}{(1+z)\left(1+\frac{z}{2}\right)\cdots\left(1+\frac{z}{n}\right)}.$$

Similarly,

$$(-z)! = \lim_{n\to\infty} \frac{n^{-z}}{(1-z)\left(1-\frac{z}{2}\right)\cdots\left(1-\frac{z}{n}\right)},$$

where, for convenience, we have used the same parameter n in both formulae. Hence,

$$(z!)(-z)! = \lim_{n\to\infty}\left[\prod_{r=1}^{n}\left(1+\frac{z}{r}\right)^{-1}\prod_{r=1}^{n}\left(1-\frac{z}{r}\right)^{-1}\right]$$

$$= \prod_{r=1}^{\infty}\left(1-\frac{z^2}{r^2}\right)^{-1}.$$

The infinite product above is convergent, and as can be shown (See Exercise 2.4), it converges to $\pi z/\sin \pi z$. Hence

$$(z!)(-z)! = \frac{\pi z}{\sin \pi z}. \tag{2.12}$$

Corollary If we put $z = -\frac{1}{2}$ in the above result, we have,

$$\left(\frac{1}{2}\right)!\left(-\frac{1}{2}\right)! = \frac{\pi/2}{\sin \pi/2} = \frac{\pi}{2}.$$

Now using $\left(\frac{1}{2}\right)! = \frac{1}{2}\left(-\frac{1}{2}\right)!$, we find

$$\left[\left(-\frac{1}{2}\right)!\right]^2 = \pi.$$

Thus,

$$\left(-\frac{1}{2}\right)! = \Gamma\left(\frac{1}{2}\right) = \sqrt{\pi}.$$

Here we have taken the positive square root since an inspection of the definition (2.8) shows that $\left(-\frac{1}{2}\right)!$ is positive.

2.3 Alternative Forms for the Gamma Function

[A] Weiterstrass definition: Weiterstrass gave a definition of $\Gamma(z)$ as

$$\frac{1}{\Gamma(z)} = ze^{\gamma z}\prod_{r=1}^{\infty}\left(1+\frac{z}{r}\right)e^{-z/r} \tag{2.13}$$

Here γ is the Euler constant[12] ($\gamma = 0.5772\ldots$). We can relate this definition to the definition of $(z)!$ given earlier. To do this, we note from (2.8),

$$(z-1)! = \Gamma(z) = \frac{z!}{z} = \lim_{n\to\infty} \frac{n!\,n^z}{z(z+1)\cdots(z+n)}$$

$$= \lim_{n\to\infty} \frac{e^{z\ln n}}{z(z+1)\left(1+\frac{z}{2}\right)\cdots\left(1+\frac{z}{n}\right)}$$

$$= \lim_{n\to\infty} \frac{1}{z} e^{z\ln n} \prod_{r=1}^{n}\left(1+\frac{z}{r}\right)^{-1}$$

$$= \lim_{n\to\infty} \frac{1}{z} e^{z\ln n} \prod_{r=1}^{n} e^{-z/r} \prod_{r=1}^{n}\left(1+\frac{z}{r}\right)^{-1} e^{z/r}$$

$$= \lim_{n\to\infty} \frac{1}{z} e^{z\left(\ln n - \sum_{r=1}^{n}\frac{1}{r}\right)} \prod_{r=1}^{n}\left(1+\frac{z}{r}\right)^{-1} e^{z/r}.$$

Taking inverse,

$$\frac{1}{\Gamma(z)} = z \lim_{n\to\infty} e^{-z\left(\ln n - \sum_{r=1}^{n}\frac{1}{r}\right)} \lim_{n\to\infty} \prod_{r=1}^{n}\left(1+\frac{z}{r}\right)e^{-z/r}.$$

Now, we know that

$$\lim_{n\to\infty}\left(\sum_{r=1}^{n}\frac{1}{r} - \ln n\right) \equiv \gamma = \text{ the Euler constant } = 0.577\ldots$$

We shall soon show the existence of the second limit. Hence, we have

$$\frac{1}{\Gamma(z)} = z e^{\gamma z} \prod_{r=1}^{\infty}\left(1+\frac{z}{r}\right)e^{-z/r},$$

thus proving the equivalence of the two definitions.

Convergence of the infinite product

Consider

$$\lim_{n\to\infty} A_n(z)$$

[12] It is possible to show that while $\sum_{r=1}^{n}\frac{1}{r}$ diverges as $n \to \infty$, the difference $\left\{\sum_{r=1}^{n}\frac{1}{r} - \ln n\right\}$ approaches a finite limit γ called the Euler constant. See e.g., *Advanced Engineering Mathematics* by Kreizig.

where,

$$A_n(z) = \prod_{r=1}^{n} \left(1 + \frac{z}{r}\right) \exp\left(-\frac{z}{r}\right).$$

Taking logarithms,

$$\ln A_n(z) = \sum_{r=1}^{n} \left\{ \ln\left(1 + \frac{z}{r}\right) - \frac{z}{r} \right\}.$$

We shall show that $\lim_{n\to\infty} \ln A_n(z)$ exists for any $z \in \mathbb{C}$. For this purpose, given any z, we split the sum as

$$\ln A_n(z) = \sum_{r=1}^{N_0} \left\{ \ln\left(1 + \frac{z}{r}\right) - \frac{z}{r} \right\} + \sum_{r=N_0+1}^{n} \left\{ \ln\left(1 + \frac{z}{r}\right) - \frac{z}{r} \right\}, \tag{A}$$

where, $N_0 = [2|z|] =$ integral part of $|2z|$, and we have assumed, without loss of generality, that $n > [2|z|]$. Now, the first term on the right-hand side has a fixed and finite number of terms (N_0) and is a finite quantity irrespective of the limit $n \to \infty$. In the second sum, we note that $|\frac{z}{r}| < 1/2 < 1$ and hence we can expand the logarithm as

$$\begin{aligned}
\left|\ln\left(1 + \frac{z}{r}\right) - \frac{z}{r}\right| &= \left| -\frac{z^2}{2r^2} + \frac{z^3}{3r^3} - \frac{z^4}{4r^4} + \cdots \right| \\
&\le \left|\frac{z^2}{2r^2}\right| + \left|\frac{z^3}{3r^3}\right| + \left|\frac{z^4}{4r^4}\right| + \cdots \\
&\le \frac{|z|^2}{2r^2} + \frac{|z|^3}{2r^3} + \frac{|z|^4}{2r^4} + \cdots \\
&= \frac{|z|^2}{2r^2} \left\{ \frac{1}{1 - \frac{|z|}{r}} \right\} \\
&\le \frac{|z|^2}{r^2}.
\end{aligned} \tag{2.14}$$

Hence, for the second term on the right-hand side of (A),

$$\begin{aligned}
\lim_{n\to\infty} \left| \sum_{r=N_0+1}^{n} \left\{ \ln\left(1 + \frac{z}{r}\right) - \frac{z}{r} \right\} \right| &\le \sum_{r=N_0+1}^{\infty} \left| \ln\left(1 + \frac{z}{r}\right) - \frac{z}{r} \right| \\
&\le \sum_{r=N_0+1}^{\infty} \frac{|z|^2}{r^2} \\
&< \infty
\end{aligned} \tag{2.15}$$

since the series $\sum_{r=1}^{\infty} \frac{1}{r^2}$ converges. Thus, the $\lim_{n\to\infty} \ln A_n(z)$ exists and equals, say $B(z)$. Thus,

$$\lim_{n\to\infty} A_n(z) = \exp B(z).$$

This proves the convergence of the infinite product in the Weiterstrass definition of $\frac{1}{\Gamma(z)}$, and also in the definitions of (2.6) and (2.8).

[B] The Euler's definition: One encounters the Γ function most commonly in the form given by Euler (or those derived from it). Euler's definition reads

$$z! = \int_0^\infty e^{-t} t^z dt \quad \mathrm{Re}\, z > -1. \tag{2.16}$$

Or equivalently,

$$\Gamma(z) = \int_0^\infty e^{-t} t^{z-1} dt. \quad \mathrm{Re}\, z > 0 \tag{2.17}$$

The real part of z is so restricted as above to ensure convergence of the integrals at the lower limit $t = 0$. (See Exercise 2.10).

[C] An alternate Euler's definition: A change of variables $t = x^2$ enables one to re-express (2.17) as

$$\Gamma(z) = 2\int_0^\infty e^{-x^2} x^{2z-1} dx. \quad \mathrm{Re}\, z > 0 \tag{2.18}$$

We shall show the equivalence of the definition of $\Gamma(z)$ by Euler and the definition we have adopted in the common domain of their definitions, viz. $\mathrm{Re}\, z > 0$. We start from $\exp(-t) = \lim_{n\to\infty} \left(1 - \frac{t}{n}\right)^n$. We replace the integral in the Euler's definition as follows:

$$\begin{aligned}\Gamma(z) &= \int_0^\infty e^{-t} t^{z-1} dt \\ &\to \lim_{n\to\infty} \int_0^n \left(1 - \frac{t}{n}\right)^n t^{z-1} dt \\ &= \lim_{n\to\infty} n^z \int_0^1 (1-\tau)^n \tau^{z-1} d\tau, \end{aligned} \tag{2.19}$$

where we have made a change of variable $\tau = t/n$ in the second step[13]. It is now possible to integrate by parts the last integral and evaluate it. We end up with

[13] The replacement in the first step calls for a careful justification that we have skipped. In this sense, the proof presented here is not completely rigorous. For a careful treatment, see e.g., *A Course on Modern Analysis* by Whittaker and Watson.

$$\Gamma(z) = \lim_{n\to\infty} n^z \frac{n(n-1)\cdots 1}{z(z+1)(z+2)\cdots(z+n-1)} \int_0^1 d\tau\, \tau^{z+n-1}$$
$$= \lim_{n\to\infty} n^z \frac{n(n-1)\cdots 1}{z(z+1)(z+2)\cdots(z+n-1)(z+n)}$$
$$= \lim_{n\to\infty} n^{z-1} \frac{n(n-1)\cdots 1}{z(z+1)(z+2)\cdots(z+n-1)} \times \lim_{n\to\infty} \frac{n}{n+z}. \tag{2.20}$$

We note that the right-hand side above just coincides with definition of (2.7) we started from, thus proving the equivalence. We note as a result that the $\Gamma(z)$ defined by (2.7) is just the analytic continuation of the Eulerian integral to the complex plane.

2.4 Digamma Function

Often, we may need the values of the first derivative of the factorial function. The digamma function is related to it.

We start from the Weiterstrass definition (2.13) of $\Gamma(z)$ and rewrite it as

$$\frac{1}{z!} = \frac{1}{z\Gamma(z)} = e^{\gamma z} \prod_{r=1}^{\infty} \left(1 + \frac{z}{r}\right) e^{-z/r}.$$

We take logarithms of both sides:

$$-\ln(z!) = \gamma z + \sum_{r=1}^{\infty} \left\{ \ln\left(1 + \frac{z}{r}\right) - z/r \right\}$$

We now differentiate with respect to z to obtain

$$F(z) \equiv \frac{d}{dz} \ln z! = \frac{1}{z!} \frac{d}{dz} z!$$
$$= -\gamma - \sum_{r=1}^{\infty} \left(\frac{1}{z+r} - \frac{1}{r} \right). \tag{2.21}$$

We note some special values:

$$F(0) = -\gamma \tag{2.22}$$

$$F(1) = -\gamma - \sum_{r=1}^{\infty} \left(\frac{1}{1+r} - \frac{1}{r} \right) = 1 - \gamma \tag{2.23}$$

$$F(n) = -\gamma + \sum_{r=1}^{n} \frac{1}{r} \tag{2.24}$$

2.5 Properties of Γ Function

In this section we shall list a number properties of the gamma function. The proofs of the statements are left as exercises (with sufficient hints given).

1. $\Gamma(z)$ never vanishes for any z. This is seen from the Weiterstrass definition of $1/\Gamma(z)$ of (2.13). We note that, as has been shown explicitly, the infinite product on the right-hand side of (2.13) converges for all z.
2. $1/\Gamma(z)$ is an entire function. This follows from the convergence of the infinite product mentioned above.
3. $\Gamma(z)$ satisfies the Schwarz reflection principle: $\Gamma^*(z) = \Gamma(z^*)$. This is easily seen from the Weiterstrass definition.

Some properties of $\Gamma(x)$ for real x

4. $\Gamma(x) = \Gamma(x + i0)$ is real. This follows from the above observation 3 or by a direct inspection of the Weiterstrass definition.
5. $\Gamma(x)$ is positive for $x > -1$. This is verified by the inspection of the Weiterstrass definition: for $x > -1$ all factors in it are positive.
6. $\Gamma(x)$ is *monotonically increasing* for all x in $(2, \infty)$. This is seen by noting that $F > 0$ for x in this interval.
7. Consider $\Gamma(x)$ for negative x as x ranges in the successive intervals $(-1, 0)$, $(-2, -1)$, $(-3, -2), \ldots$. We find that $\Gamma(x)$ alternates in sign from one interval to the next. This is seen by the use of property 5 above and the relation $\Gamma(x) = \frac{\pi}{\sin \pi x \cdot \Gamma(1-x)}$. See also Exercise 2.11.
8. The asymptotic behaviour of $\Gamma(x)$ for large real x has been obtained in (*Mathematical Physics: The Basics*) volume 1 (see section 8.3 and Exercise 8.13) and leads to what is called the Sterling approximation:

$$\Gamma(x+1) = e^{-x} x^{x+\frac{1}{2}} \sqrt{2\pi} \left\{ 1 + \frac{1}{12x} + O\left(\frac{1}{x^2}\right) \right\}$$

2.6 Beta Function

Beta function occurs during the evaluation of several classes of definite integrals, as already mentioned in the introduction to this chapter. To recall some of these,

$$\int_0^{\pi/2} \sin^m \theta \cos^n \theta d\theta;$$

$$\int_0^1 x^m (1-x)^n dx;$$

$$\int_0^\infty \frac{x^m}{(1+x)^n} dx;$$

(each under suitable conditions on m and n). Moreover, as we shall soon see, the beta function is related to the gamma function through the relation

$$B(m,n) = \frac{\Gamma(m)\Gamma(n)}{\Gamma(m+n)}. \tag{2.25}$$

We evaluate the class of integrals above and the related ones by their connection to Γ-function given above.

Definition: $B(m,n)$ is defined for any complex m and n (subject to the restriction $\mathrm{Re}\, m > 0$, $\mathrm{Re}\, n > 0$) by the definite integral

$$B(m,n) = \int_0^1 dt\, t^{m-1}(1-t)^{n-1}. \tag{2.26}$$

Alternate forms of the beta function, $B(m,n)$, can be obtained by a variety of substitutions in (i). For example,

1. $t = \sin^2\theta$ leads to

$$B(m,n) = 2\int_0^{\pi/2} (\sin\theta)^{2m-1}(\cos\theta)^{2n-1} d\theta. \tag{2.27}$$

2. $t = \frac{1}{1+u}$, with $(1-t) = \frac{u}{(1+u)}$, leads to

$$B(m,n) = \int_0^\infty \frac{u^{(n-1)}du}{(1+u)^{m+n}}. \tag{2.28}$$

3. $t = \frac{1}{u}$, with $(1-t) = \left(\frac{u-1}{u}\right)$ leads to

$$B(m,n) = \int_1^\infty du \frac{(u-1)^{n-1}}{u^{m+n}}. \tag{2.29}$$

2.7 Relation between Beta and Gamma Functions

As mentioned in the previous section, a relation between the beta and the gamma functions exists as given by (2.25). It is established by the use of the following representations for Γ and B functions:

$$\Gamma(m) = 2\int_0^\infty dx e^{-x^2} x^{2m-1} \quad \mathrm{Re}\, m > 0$$

and

$$B(m,n) = 2\int_0^{\pi/2} (\sin\theta)^{2n-1}(\cos\theta)^{2m-1} d\theta \quad \mathrm{Re}\, m, \mathrm{Re}\, n > 0.$$

To prove the result, we write down, in addition to the above,

$$\Gamma(n) = 2\int_0^\infty dye^{-y^2}y^{2n-1}dy \quad \mathrm{Re}\, n > 0,$$

where we have chosen the integration variable as y. We now consider the product

$$\Gamma(m)\Gamma(n) = 4\int_0^\infty dxe^{-x^2}x^{2m-1}dx \times \int_0^\infty dye^{-y^2}y^{2n-1} \quad \mathrm{Re}\, m, \mathrm{Re}\, n > 0.$$

Now, we shall look upon x and y, the arbitrary integration variables, as the Cartesian coordinates. Then the above product can be looked upon as an integral over an *area* in the x–y plane

$$= 4\int\int dxdye^{-(x^2+y^2)}x^{2m-1}y^{2n-1},$$

where, on account of the limits $0 \le x, y \le \infty$ on the variables, the area integral goes over the area of the *first* quadrant.

We may then re-evaluate the above integral using the polar coordinates (r, θ). We recall the alternate way of writing the area element:

$$dA = dxdy \to rdrd\theta \quad 0 \le r < \infty \quad 0 \le \theta \le \pi/2$$

We, thus, have

$$\Gamma(m)\Gamma(n) = 4\int_0^\infty rdr\int_0^{\pi/2} d\theta e^{-r^2}r^{2m+2n-2}(\cos\theta)^{2m-1}(\sin\theta)^{2n-1}.$$

The integral on the right-hand side is factorisable. We thus have

$$\Gamma(m)\Gamma(n) = 2\int_0^\infty drr^{2m+2n-1}e^{-r^2} \times 2\int_0^{\pi/2} d\theta(\cos\theta)^{2m-1}(\sin\theta)^{2n-1}$$

$$\mathrm{Re}\, m, \mathrm{Re}\, n > 0.$$

We recognise, from the representations for Γ and B of (2.18) and (2.27), the right-hand side equals $\Gamma(m+n)B(m,n)$.

Thus,

$$B(m,n) = \frac{\Gamma(m)\Gamma(n)}{\Gamma(m+n)}.$$

This proves the result.

Solved problem 2.1

(a) Evaluate

$$I = \int_0^{\pi/2} d\theta (\cos\theta)^{2m-1} (\sin\theta)^{2n-1},$$

where m and n are positive integers.

(b) Obtain the value for $m = n = 2$

(c) Find the limit $n \to 1$ of I in (a).

Solution:

(a) We note from (2.27):

$$I = \frac{1}{2} B(m, n) = \frac{1}{2} \frac{\Gamma(m)\Gamma(n)}{\Gamma(m+n)} = \frac{1}{2} \frac{(m-1)!(n-1)!}{(m+n-1)!}$$

(b) For $m = n = 2$,

$$I = \frac{1}{2} \frac{1}{3!} = \frac{1}{12}.$$

(c)
$$I = \frac{1}{2} \frac{(m-1)!(n-1)!}{(m+n-1)!}$$

As $n \to 1$,

$$I \to \frac{1}{2} \frac{(m-1)!0!}{m!} = \frac{1}{2} \frac{1}{m} \qquad \text{Re}\, m > 0.$$

Exercises

2.1 Find the limit $\epsilon \to 0$ of $\frac{\Gamma(-1+\epsilon)\Gamma(\epsilon/2)}{[\Gamma(\epsilon)]^2}$.

2.2 Find the region of the complex z-plane for which the integral $\int_0^\infty dt\, t^z \exp(-t^4)$ is defined. Give its analytic continuation over the entire complex plane (except possibly for isolated singularities).

2.3 Show that $\lim_{\epsilon \to 0} \frac{\Gamma(-1+2\epsilon)}{\Gamma(\epsilon)} = -1/2$.

2.4 Consider the function $f(z) = \cot z - 1/z$. Let C_r be a square contour with centre at the origin and sides parallel to axes and of length $= (2r+1)\pi$.

(a) Apply the residue theorem to evaluate

$$\oint_{C_r} dz \left[\frac{\cot z - \frac{1}{z}}{z - z_0} \right],$$

where z_0 is a point within C_r.

(b) Show that the integral over C_r goes to zero as $r \to \infty$. Hence obtain $f(z_0)$.
(c) Let $g(z) = \frac{1}{z}\sin z$. Show that $f(z) = \frac{g'(z)}{g(z)} = \frac{d}{dz}\ln g(z)$. Integrate $\int_0^z dz' f(z')$ using the result in (b).
(d) Hence show that $\frac{\sin z}{z} = \prod_{r=1}^{\infty}\left[1 - \frac{z^2}{\pi^2 r^2}\right]$.

2.5 Verify that the Sterling approximation $N! \propto e^{-N}N^{N+\frac{1}{2}}$ is consistent with $(N+1)! = (N+1)N!$ to the leading order.
2.6 Use the Sterling approximation for $(2N)!$ and $N!$ in the duplication formula to arrive at the leading approximation for $\left(N - \frac{1}{2}\right)!$ i.e., for large half-odd-integers.

On § 2.3

2.7 Show that

$$\left|\frac{\Gamma(x+iy)}{\Gamma(x)}\right| = \prod_{r=0}^{\infty}\left[1 + \frac{y^2}{(x+r)^2}\right]^{-1/2}.$$

Hence show that $|\Gamma(x+iy)| \le \Gamma(x)$.
2.8 Show that

$$(a) \int_0^1 dx\, x^k \ln x = \frac{1}{(k+1)^2} \qquad k > -1.$$

2.9 Consider the integral $I(z) = \int_0^\infty dt \exp(it)t^{z-1}$. Show that for a certain domain in the complex z-plane, I can be expressed as $I(z) = i^z\Gamma(z)$. Find this domain. Give analytic continuation of $I(z)$ to the complex plane and specify its domain of definition. [Hint: You may try a suitable contour and Jordan lemma.]
2.10 Show that the definition of $\Gamma(z)$ of (2.17) is meaningful for $\text{Re}\, z > 0$ by analyzing the convergence of the integral near the lower end at $t \sim 0$.

On § 2.5

2.11 Consider $\Gamma(x)$ (x real) for successive intervals of the kind $(-n-1, -n)$ where n is a nonnegative integer. Show that the sign of $\Gamma(x)$ alternates as one goes from such an interval to the next. [Hint: $\Gamma(x+1) = x\Gamma(x)$.]
2.12 Show that $\Gamma(x)$ is a monotonically decreasing function of x for $0 < x < 1$ and a monotonically increasing function for $2 < x < \infty$. [Hint: Consider $\frac{1}{\Gamma(x)}\frac{d\Gamma}{dx}$ and show that this is respectively negative and positive.]
2.13 Consider the function $\Gamma(x)$ for $3 < x < 4$. Plot the function in this interval. Further, using

$$\Gamma(-x) = \frac{-\pi}{\sin \pi x \Gamma(x+1)},$$

plot the function for $-3 < x < -2$. Show that $\Gamma(x)$ has one maximum at $x_m \in (-3, -2)$. Find if $x_m > -2.5$. Use the above approach to plot the gamma function for all real x; $-\infty < x < \infty$.

2.14 Study the plot of $\Gamma(x)$ for x real. Hence argue that $\Gamma(x)$ never vanishes for $-\infty < x < \infty$.

2.15 Show that the equation $x! = k, k \neq 0$, has an infinite number of real roots.

2.16 Show that $\Gamma(z)$ satisfies the Schwarz reflection principle. Verify that $\Gamma(x)$ is real on the real axis. Show that $|\Gamma(z)|^2 = \Gamma(z)\Gamma(z^*)$.

2.17 Show that $\frac{1}{\Gamma(z)}$ is an entire function. Using the Liuville's theorem argue that there are points in the complex plane where $\Gamma(z)$ approaches arbitrarily close to zero but does not vanish anywhere. [Hint: Recall Weiterstrass definition.]

2.18 Show that $\Gamma(x)$ is positive for $x > 0$.

On § 2.6

2.19 The following occurs in connection with calculation in Quantum Field Theory:

$$I(z, \alpha, \beta) = \int_0^\infty \frac{r^{z-1} dr}{(r^2 + m^2)^\alpha r^{2\beta}}$$

where α and β are nonnegative integers and z is a complex variable.

(a) Under what restrictions on z, α and β does this integral exists?

(b) What is the analytical continuation, $\tilde{I}(z, \alpha, \beta)$, of the above integral for the entire complex plane (barring isolated singularities)?

(c) Find

$$\lim_{z \to 4} \{(z-4)\tilde{I}(z, \alpha, \beta)\}.$$

2.20 For what values of α and β does $\tilde{I}(z, \alpha, \beta)$ in the previous problem have a simple pole, a double pole, a removable singularity and no pole at $z = 4$?

On § 2.7

2.21 Consider the following integral $I(p, q) = \int_0^{\pi/2} d\theta (\sin\theta)^p (\cos\theta)^q$. (a) Evaluate it for $p = 4 = q$. (b) Suppose we look upon $f(p) \equiv I(p, q = p)$ as a function of the *complex* variable p. (i) Write the domain of definition of $f(p)$. (ii) Write the analytic continuation of $f(p)$ to the complex p-plane and specify its domain of definition. Evaluate it for $p = -6$.

2.22 Let $I(a) = \int_0^{\pi/2} d\theta (\sin\theta)^{(1+2a)} (\cos\theta)^{(1-2a)}$, where a is a real parameter with $|a| < 1$. Find $\frac{dI(a)}{da}\Big|_{a=0}$.

2.23 Show that

$$\int_0^{\pi/2} d\theta (\sin\theta)^{\frac{5}{2}} (\cos\theta)^{\frac{-1}{2}} = \frac{3\pi}{4\sqrt{2}}.$$

2.24 For what values of m and n can the evaluation of the integral $\int_0^{\pi/2} d\theta (\sin\theta)^m (\cos\theta)^n$ be done by applying the residue theorem to a contour integral over a unit circle? Perform the evaluation for the case $m = n = 2p$, (p is a positive integer) and compare with the result obtained from the use of the beta function.

2.25 Express $\int_0^{\pi/2} d\theta (\sin\theta)^{\frac{3}{4}} (\cos\theta)^{\frac{-1}{4}}$ in terms only of $\Gamma\left(\frac{1}{4}\right)$, π and real numbers.

Additional problems

2.26 Show that the function $F(z)$ such that (i) $F(n) = n!$ ($n = 0, 1, 2, \ldots$) (ii) $F(z)$ is meromorphic is not sufficient to uniquely fix $F(z)$. Here, we note that a meromorphic function is one whose only singularities are the poles. [Hint: Consider $\tilde{F}(z) = F(z) + A \sin \pi z$.]

2.27 In the previous problem F is specified on a set of infinite number of points $z = n$ (n is a nonnegative integer). Do the results on analytic continuation, then imply the uniqueness of the analytic continuation $F(z)$. If not, why not?

2.28 Suppose $F\left(\frac{1}{z}\right)$ is specified as

$$F\left(\frac{1}{z}\right) = n! \quad z = \frac{1}{n} \quad n = 1, 2, \ldots$$

This specifies $G(z) = F\left(\frac{1}{z}\right)$ on a point set $\{z = \frac{1}{n}\ n = 1, 2, \ldots\}$ having a limit point. What is the limit point? Do the results on the uniqueness of analytic continuation now imply that $G(z)$ is unique? Discuss.

References

[1] Whittaker, E. T. and Watson, G. N., *A Course of Modern Analysis: An Introduction to the General Theory of Infinite Processes and of Analytic Functions, with an Account of the Principal Transcendental Functions*, AMS Press, New York, 1979.

3 Bessel Functions

3.1 Introduction

Among the special functions commonly occurring in Physics, there are two important ones: (i) the Bessel functions (and the related ones) and (ii) the Legendre Polynomials. In the present chapter, we shall study Bessel functions of various kinds. To be specific, we shall study the cylindrical Bessel functions $J_\nu(x)$, the spherical Bessel functions $j_n(x)$, the cylindrical Neumann functions $N_\nu(x)$, the spherical Neumann functions $n_n(x)$ [as also certain specific linear combinations of these called the Hankel functions (cylindrical and spherical) of two kinds: $H_\nu^{(1,2)}(x)$ and $h_\nu^{(1,2)}(x)$]. We shall also briefly study Bessel functions with imaginary arguments; these are related to the so called modified Bessel functions.

It is perhaps worthwhile to dwell upon how and why these Bessel functions come to be so conspicuous in Physics. It turns out that this rests upon the fact that several of the physical laws are expressed as the second order differential equations involving the Laplacian ∇^2. Among the various physical laws that you may be familiar with (such as the Maxwell's equations, equation for the Newtonian gravitational potential, the Schrödinger equation in quantum mechanics, etc.) as well as those as yet unfamiliar (relativistic wave equations, equations of general relativity, etc.), two of the equations that commonly occur are the Poisson equation

$$\nabla^2\phi = -\rho \tag{3.1}$$

[including the Laplace's equation ($\nabla^2\phi = 0$)] and the Helmholtz wave equation

$$(\nabla^2 + k^2)\psi = 0. \tag{3.2}$$

The Poisson/Laplace's equations occur in electrostatics/electrodynamics and Newtonian gravity. The Helmholtz wave equation occurs when the wave equations (whether for electromagnetic waves, sound waves or other relativistic fields) are cast in a time-independent form. It also appears in the time-independent Schrödinger equation for a free particle.

Thus, in the wave equation

$$\left(\nabla^2 - \frac{1}{c^2}\frac{\partial^2}{\partial t^2}\right)\phi(\mathbf{x}, t) = 0 \tag{3.3}$$

a substitution of $\phi(\mathbf{x}, t) = e^{-i\omega t}\phi(\mathbf{x})$ (representing a monochromatic wave) leads one to

$$(\nabla^2 + k^2)\phi(\mathbf{x}, t) = 0, \quad k^2 \equiv \omega^2/c^2 \tag{3.4}$$

whereas in the time-dependent Schrödinger equation for a free particle

$$-i\hbar\frac{\partial\psi}{\partial t} = -\frac{\hbar^2}{2m}\nabla^2\psi \tag{3.5}$$

the substitution of $\psi(\mathbf{r}, t) = e^{-iEt/\hbar}\psi(\mathbf{r})$ (corresponding to energy eigenfunctions) leads to

$$(\nabla^2 + k^2)\psi(\mathbf{r}) = 0 \quad k^2 \equiv \frac{2mE}{\hbar^2}.$$

It is easy to show that (see e.g., Exercises 3.7 and 3.14):

1. When the Laplace's equation is solved by the method of separation of variables in the cylindrical coordinates, one ends up with the Bessel equation (modified Bessel equation) for the radial part. Such solutions are relevant in electrostatics.
2. When the Helmholtz wave equation is solved by the method of separation of variables in the spherical polar coordinates, the radial equation is the spherical Bessel equation.

In addition, Bessel functions occur for example in diffraction problems, and vibrations of membranes which are related to wave phenomena involving cylindrical objects. The spherical Bessel functions are relevant in the multipole expansions of fields, in the quantum mechanical solution to the bound state problem in a square well (the deuteron problem) and the scattering problem from a square well/barrier. In a addition, the spherical Bessel functions, $j_l(kr)$ naturally occur in the 'partial wave expansion' of a plane wave in terms of the spherical waves

$$e^{ikr\cos\theta} = \sum_{l=0}^{\infty}(2l+1)\,j_l(kr)i^l P_l(\cos\theta). \tag{3.6}$$

This expansion is, in particular, useful in the scattering problems involving all spherically symmetric potentials. The occurrence of $j_l(kr)$ in the above relation is related to the remark (2) above – that both sides of (3.6) are solutions of the Helmholtz wave equation (3.4). The scattering problems in quantum mechanics is a natural place to employ spherical functions $h_l^{(1,2)}(x) = j_l(x) \pm in_l(x)$ on account of their asymptotic nature: they represent the outgoing and the incoming waves.

As the above examples illustrate, the Bessel functions $J_\nu(x)$ occur in problems of the above type with cylindrical symmetry and as such are also called the cylindrical Bessel functions. (When we ordinarily refer to Bessel functions, we normally mean *cylindrical* Bessel functions.) On the other hand, problems of the above type with *spherical* symmetry involve spherical Bessel functions. These Bessel functions satisfy a certain second order ODE of the kind dealt with in Chapter 1. As discussed in Chapter 1, such a second order ODE has two linearly independent solutions. A choice of such as linearly independent solutions is often made as the pair $(J_\nu(x), N_\nu(x))$ for the cylindrical Bessel equation, and as $(j_\nu(x), n_\nu(x))$ for the spherical Bessel equation, $N_\nu(x)$ and $n_\nu(x)$ being the Neumann functions of cylindrical/spherical types. Neumann function can occur in solutions to the class of problems listed above in regions that exclude $\rho = 0$ or $r = 0$. Asymptotic behaviour (as $\rho \to \infty$ or $r \to \infty$) makes it often more useful to employ their (complex) linear combinations called Hankel functions $H_\nu^{(1,2)}(x) = J_\nu(x) \pm iN_\nu(x)$; $h_\nu^{(1,2)}(x) = j_\nu(x) \pm n_\nu(x)$.

3.2 Bessel Functions: Definition

A Bessel function can be defined in a number of ways: either as a convergent series or as a contour integral. We shall choose to adopt the series definition and later discuss briefly the contour integral expressions.

Definition: For a real ν, the (cylindrical) Bessel function $J_\nu(x)$ of order ν and with the argument x is defined by the convergent series

$$J_\nu(x) = \sum_{r=0}^{\infty} \frac{(-)^r}{r!(\nu+r)!} \left(\frac{x}{2}\right)^{\nu+2r}. \tag{3.7}$$

We note that:

1. For nonintegral ν, the above definition contains (in the denominator) a factorial of $(\nu + r)$, which is a noninteger. Such a factorial has been defined in Chapter 2.
2. The above series is uniformly and absolutely convergent for all x as is seen from the ratio test of convergence (see Exercise 3.1 and also the solved problem below).
3. We shall primarily be interested in Bessel functions with its argument x *real*.

To obtain familiarity with the series (3.7) we shall study some properties of the series for $J_{1/2}$.

Solved problem 3.1

(a) Show that the series for $J_{1/2}$ converges absolutely.
(b) Study the behaviour of $J_{1/2}$ near $x \approx 0$.
(c) Show how $J_{1/2}$ is related to a sinusoidal function.
(d) Find the upper bound on $\left|J_{\frac{1}{2}}(x)\right|$ for x real.

Solution:

(a) From its definition,

$$J_{\frac{1}{2}}(x) = \sum \frac{(-)^r}{r!(r+\frac{1}{2})!} \left(\frac{x}{2}\right)^{\left(\frac{1}{2}+2r\right)}.$$

The ratio of successive terms in the series for $J_{1/2}$ is

$$\frac{a_{r+1}}{a_r} = \frac{-(x/2)^2}{(r+1)(r+3/2)} \to 0$$

as $r \to \infty$ for any x. Thus,

$$\left|\frac{a_{r+1}}{a_r}\right| \to 0 \text{ as } r \to \infty$$

and hence the series converges absolutely for all x.

(b) Near $x \approx 0$, the lowest power of x obtained by setting $r = 0$ dominates. Hence,

$$J_{1/2}(x) \approx \frac{\sqrt{x/2}}{(1/2)!} = \sqrt{\frac{2x}{\pi}}.$$

(c) Recall the duplication formula

$$r!\left(r - \frac{1}{2}\right)! = (2r)!2^{-2r}\sqrt{\pi}.$$

Thus,

$$\begin{aligned} J_{\frac{1}{2}}(x) &= \frac{1}{\sqrt{\pi}} \sum_{r=0}^{\infty} \frac{(-)^r \sqrt{2}\, x^{\left(\frac{1}{2}+2r\right)}}{(2r)!(2r+1)} \\ &= \sqrt{\frac{2}{x\pi}} \sum_{r=0}^{\infty} (-)^r \frac{x^{2r+1}}{(2r+1)!} \\ &= \sqrt{\frac{2}{\pi x}} \sin x = \sqrt{\frac{2}{\pi}} \frac{\sin x}{\sqrt{x}}. \end{aligned}$$

We note that $\left|J_{\frac{1}{2}}(x)\right|$ is simply related to a sinusoidal function. (As seen in Section 3.8, $J_{n+\frac{1}{2}}$ can be related for all integral n to sinusoidal functions.)

(d) A simple exercise shows that $\sin x/\sqrt{x}$ is maximum whenever $\tan x = 2x$. The smallest value of x for which this happens is $x \approx 1.066$ radians. At this point $J_{\frac{1}{2}}(x) \approx 0.788\ldots$ So,

$$\left|J_{\frac{1}{2}}(x)\right| \le 0.788.$$

3.3 Recurrence Relations

From the definition given above, we can derive a number of properties of the Bessel functions. Two of these are in the form of recurrence relations that relate a Bessel function of order ν (together with their first derivatives) to ones of order $(\nu \pm 1)$. The derivations are straightforward, as seen below.

Recurrence relation I

$$\frac{d}{dx}(x^\nu J_\nu) = x^\nu J_{\nu-1}(x)$$

Proof: Consider

$$\frac{d}{dx}(x^\nu J_\nu) = \frac{d}{dx}\sum_{r=0}^{\infty} \frac{(-)^r}{r!(\nu+r)!} \frac{x^{2\nu+2r}}{2^{\nu+2r}}.$$

In view of the uniform convergence of the series for $J_\nu(x)$, we can differentiate the series term by term. We thus obtain

$$\frac{d}{dx}(x^\nu J_\nu) = \sum_{r=0}^{\infty} \frac{(-)^r}{r!(\nu+r-1)!} \frac{x^{2\nu+2r-1}}{2^{\nu+2r-1}}$$

$$= x^\nu \sum_{r=0}^{\infty} \frac{(-)^r}{r![(\nu-1)+r]!} \left(\frac{x}{2}\right)^{(\nu-1)+2r}$$

$$= x^\nu J_{\nu-1}(x).$$

Recurrence relation II

$$\frac{d}{dx}(x^{-\nu} J_\nu) = -x^{-\nu} J_{\nu+1}$$

The proof of this relation also proceeds along the same lines. Consider

$$\frac{d}{dx}(x^{-\nu} J_\nu) = \frac{d}{dx}\left[\sum_{r=0}^{\infty} \frac{(-)^r}{r!(\nu+r)!} \frac{x^{2r}}{2^{\nu+2r}}\right].$$

Note that the first term $r = 0$, being independent of x, does not contribute. Hence

$$\frac{d}{dx}(x^{-\nu} J_\nu) = \sum_{r=1}^{\infty} \frac{(-)^r}{(r-1)!(\nu+r)!} \frac{x^{2r-1}}{2^{\nu+2r-1}}.$$

We change the summation index r to $r = p + 1$. Then p goes from 0 to ∞. Hence

$$\frac{d}{dx}(x^{-\nu} J_\nu) = \sum_{p=0}^{\infty} \frac{(-)^{p+1}}{p!(\nu+1+p)!} \frac{x^{2p+1}}{2^{\nu+2p+1}}$$

$$= -x^{-\nu} \sum_{p=0}^{\infty} \frac{(-)^p}{p!(\nu+1+p)!} \left(\frac{x}{2}\right)^{\nu+1+2p}$$

$$= -x^{-\nu} J_{\nu+1}(x).$$

The recurrence relations I and II allow us to obtain $J_{\nu\pm1}$ in terms of $J_\nu(x)$ and its first order derivative. They are:

$$J_{\nu-1} = x^{-\nu} \frac{d}{dx}[x^\nu J_\nu] \tag{I}$$

$$J_{\nu+1} = -x^{\nu} \frac{d}{dx}[x^{-\nu} J_\nu] \tag{II}$$

These two recurrence relations can be also written in several different useful forms either by themselves or by combining them together.

Thus, on explicit differentiation, I and II above read:

$$J_{\nu-1} = J_\nu' + \frac{\nu}{x} J_\nu \tag{III}$$

$$J_{\nu+1} = -J_\nu' + \frac{\nu}{x} J_\nu \tag{IV}$$

Adding III and IV, we obtain

$$2\frac{\nu}{x} J_\nu = J_{\nu-1} + J_{\nu+1}. \tag{V}$$

Subtracting IV from III, we obtain

$$2J_\nu' = J_{\nu-1} - J_{\nu+1}. \tag{VI}$$

A further familiarity with the Bessel series and the recurrence relations is obtained with the help of the following example.

Solved problem 3.2 Show that

1. J_0 is even and that J_1 is odd.
2. $J_0' = -J_1$
3. Between two successive zeros of J_0, there is at least one zero of J_1.

4. $\lim_{x\to 0} J_0 = 1; \quad \lim_{x\to 0} J_1 = 0.$
5. Assuming that $J_0(x) \to 0$ as $x \to \infty$ (see Section 3.9), show that

$$\int_0^\infty J_1 dx = 1.$$

Solution:

1. The series for $J_0(x)$ has even powers of x and that for $J_1(x)$ has odd powers of x.
2. We put $\nu = 0$ in the recurrence relation

$$J_{\nu+1} = -J'_\nu + \frac{\nu}{x} J_\nu$$

to obtain $J_1 = -J'_0$.

3. Let x_m and x_{m+1} be the mth and $(m+1)^{\text{th}}$ zero of J_0. Then by Rolle's theorem, J'_0 vanishes at some point between x_m and x_{m+1}. Using the result in (b) above, J_1 must vanish there. (The discussion in Section 3.10.3 also further shows that there is exactly one such zero of J_1.)
4. Set $x = 0$ in $J_\nu(x)$ for $\nu = 0$ and $\nu = 1$.
5. $\int_0^\infty J_1(x) = -\int_0^\infty J'_0(x)dx = -J_0(\infty) + J_0(0) = 1$, given $J_0(\infty) = 0$.

3.4 Differential Equation Satisfied by $J_\nu(x)$

We have already seen in Chapter 1 that the function J_ν, defined by the series (3.7), is in fact a solution of the differential equation called the Bessel's equation

$$J''_\nu + \frac{1}{x} J'_\nu + \left(1 - \frac{\nu^2}{x^2}\right) J_\nu = 0. \tag{3.8}$$

This can, however, also be derived directly from the recurrence relations (III) and (IV) as shown below. We first differentiate (III):

$$J'_{\nu-1} = J''_\nu + \frac{\nu}{x} J'_\nu - \frac{\nu}{x^2} J_\nu. \tag{3.9}$$

We next let $\nu \to \nu - 1$ in (IV) and rearrange the terms:

$$J'_{\nu-1} = -J_\nu + \frac{(\nu-1)}{x} J_{\nu-1} \tag{3.10}$$

We then substitute for $J_{\nu-1}$ from (III) in the above.

$$J'_{\nu-1} = -J_\nu + \frac{(\nu-1)}{x}\left(J'_\nu + \frac{\nu}{x}J_\nu\right) \tag{3.11}$$

Comparing the expression for $J'_{\nu-1}$ from (3.9) and (3.11) and rearranging terms one obtains (3.8).

Note that the above differential equation (3.8) contains only $\nu^2 = (-\nu)^2$. Thus, $J_{-\nu}$ will satisfy exactly the same differential equation. Thus J_ν and $J_{-\nu}$ both satisfy the same Bessel's equation and are therefore solutions of it. Are they linearity independent solutions?

To this end we recall from Chapter 1 that if two functions are linearly dependent, then they are *proportional*. In other words, J_ν and $J_{-\nu}$ are linearly independent unless $\frac{J_\nu}{J_{-\nu}}$ = a constant. It is easily seen that this is not the case when $\nu \neq$ an integer by considering the behaviour of each for small x which is as x^ν and $x^{-\nu}$ respectively. Thus the two solutions, J_ν and $J_{-\nu}$, are linearly independent solutions of the Bessel equation. On the other hand, when $\nu = n$, a *positive* integer, $J_{-\nu}$ is proportional to J_ν as seen below.

$$J_{-n}(x) = \sum_{r=0}^{\infty} \frac{(-)^r}{r!(-n+r)!}\left(\frac{x}{2}\right)^{-n+2r}$$

Now note that $(-n+r)!$ is infinite for $r = 0, 1, \ldots, n-1$, being the factorial of a negative integer. Hence the first n terms in the above expansion vanish. We then have

$$J_{-n}(x) = \sum_{r=n}^{\infty} \frac{(-)^r}{r!(-n+r)!}\left(\frac{x}{2}\right)^{-n+2r}.$$

We set $r = p + n$. The summation index 'p' now varies from zero to infinity. Thus, we can write

$$J_{-n}(x) = \sum_{p=0}^{\infty} \frac{(-)^{n+p}}{(p+n)!p!}\left(\frac{x}{2}\right)^{n+2p} \tag{3.12}$$

$$= (-)^n J_n(x). \tag{3.13}$$

This gives us the required relation.

Solved problem 3.3 Verify *directly from the definition* that $J_{\frac{1}{2}}$ and $J_{-\frac{1}{2}}$ are linearly independent solutions of the Bessel equation with $\nu = \frac{1}{2}$.

Solution: As seen above, $J_{\pm\frac{1}{2}}$ are solutions of the Bessel equation with $\nu = \frac{1}{2}$. To show the linear independence of these two solutions directly, we consider the relation

$$aJ_{\frac{1}{2}}(x) + bJ_{-\frac{1}{2}}(x) = 0, \tag{i}$$

i.e.,

$$ax^{\frac{1}{2}} \sum \frac{(-)^r}{r!\left(r+\frac{1}{2}\right)!}\left(\frac{x}{2}\right)^{2r} + bx^{-\frac{1}{2}} \sum \frac{(-)^r}{r!\left(r-\frac{1}{2}\right)!}\left(\frac{x}{2}\right)^{2r} = 0. \tag{ii}$$

Now, multiply (ii) by $x^{1/2}$ and let $x \to 0$. We find

$$b = 0. \tag{iii}$$

Put (iii) in (ii), divide by $x^{\frac{1}{2}}$ and set $x = 0$. We find

$$a = 0. \tag{iv}$$

(iii) and (iv) imply the linear independence of $J_{\pm\frac{1}{2}}(x)$.

(The above argument can intuitively be understood by realizing that $J_{\frac{1}{2}}$ and $J_{-\frac{1}{2}}$ behave differently near $x = 0$, viz., as $x^{\pm\frac{1}{2}}$, and hence cannot be linearly dependent. Equivalently, two functions which are linearly dependent are proportional, and the distinct behaviour of $J_{\frac{1}{2}}$ and $J_{-\frac{1}{2}}$ near $x = 0$ implies that this impossible.)

3.5 The Generating Function for Bessel Functions of an Integral Order

We would first like to introduce a new idea, very useful, in dealing with several types of special functions, viz., that of a *generating function*. For example, we may consider, at a time, the class of (i) all Bessel functions of integral orders $\{J_n : n = 0, \pm 1, \pm 2, \ldots\}$, or (ii) all Legendre polynomials $\{P_l(x) : l = 0, 1, 2, \ldots\}$, or (iii) all Hermite polynomials $\{H_n(x) : n = 0, 1, 2, \ldots\}$, etc. All functions in each of these classes have several properties in common. For example, for the class $\{J_n(x) : n = 0, \pm 1, \pm 2, \ldots\}$, each of $J_n(x)$ shares the same form of recurrence relations, the same form of integral representations, etc.

It is possible to deal simultaneously with an entire class of special functions with the help of an appropriate generating function. Thus, the generating function for Bessel functions $J_n(x)$ of integral orders simultaneously contains in it information about of these Bessel functions. The same applies to, say, the generating function for the Legendre and Hermite polynomials.

The generating function thus enables us to deal with the entire class simultaneously and allows one to derive results applicable to all of them at once. It will also enable us to further obtain an integral representation for Bessel functions of integral order n.

We shall now show that

$$g(t) = \exp\left[\frac{x}{2}\left(t - \frac{1}{t}\right)\right] \tag{3.14}$$

serves as a generating function for Bessel functions of integral orders, in the sense that each of the $J_n(x)$'s can be recovered from it. $g(t)$ will be shown to have an expansion

$$g(t) = \sum_{n=-\infty}^{\infty} t^n J_n(x) \tag{3.15}$$

valid for any t, real or complex. Thus, $J_n(x)$ can then be read off as the coefficient of t^n in the (Laurent-like) expansion of $g(t)$ in both the (positive and negative) powers of t. We shall now demonstrate this result.

Proof: The proof proceeds in a very straightforward manner: We expand out $g(t)$ in powers of t and simply collect together the coefficient of t^n. We then compare this coefficient with the series definition of $J_n(x)$ and recognise its identity.

First, we expand out the exponential as a power series.

$$g(t) = \exp\left[\frac{x}{2}\left(t - \frac{1}{t}\right)\right] = \sum_{r=0}^{\infty}\left(\frac{x}{2}\right)^r \left(t - \frac{1}{t}\right)^r \frac{1}{r!} \tag{3.16}$$

We next expand $\left(t - \frac{1}{t}\right)^r$ using the binomial theorem

$$\begin{aligned}\left(t - \frac{1}{t}\right)^r &= \sum_{p=0}^{r} \frac{r!}{p!(r-p)!} t^{r-p} \left(\frac{-1}{t}\right)^p \\ &= \sum_{p=0}^{r} \frac{r!(-)^p}{p!(r-p)!} t^{r-2p}.\end{aligned} \tag{3.17}$$

Substituting this in (3.16) above, we obtain

$$g(t) = \sum_{r=0}^{\infty}\left(\frac{x}{2}\right)^r \sum_{p=0}^{r} \frac{(-)^p t^{r-2p}}{p!(r-p)!} \tag{3.18}$$

This is a double series and we shall assume that it can be rearranged. We shall then collect the coefficient of a given power of t. To do this, we set in the above, $r - 2p \equiv n$, a fixed number. Thus, from the double summation only those terms with $r - 2p = n$ will contribute and one will be left with a *single* summation in the coefficient of t^n. We take p as the summation variable. Then r becomes fixed at $r = 2p + n$. We shall first choose $n \geq 0$. Now, in the above double summation, p will be allowed to vary from 0 to ∞. (This is so because in the original double summation, r can be taken as large as required.) The coefficient of t^n in the above double summation is then

$$\sum_{p=0}^{\infty} \left(\frac{x}{2}\right)^{n+2p} \frac{(-1)^p}{p!(n+p)!} \tag{3.19}$$

But this is precisely the definition of $J_n(x)$. This thus shows that the coefficients of the nonnegative powers of t are in fact the Bessel functions of the respective orders.

We shall next consider the case $n < 0$. Since $r = 2p + n$ and $(r - p) \geq 0$ in (3.18); we have $p + n \geq 0$, i.e., $p \geq -n = |n|$. The coefficient of t^n in (3.18) then is

$$\sum_{p=|n|}^{\infty} \left(\frac{x}{2}\right)^{n+2p} \frac{(-1)^p}{p!(n+p)!} \tag{3.20}$$

In the above equation, we can now enlarge the range of summation index from $p = 0$ to ∞. This is so because, in doing so, we add terms proportional to $\frac{1}{(n+p)!}$; $0 \leq p \leq [|n| - 1]$, and all of these terms vanish since they involve the inverse of the factorial of a negative integer. When the range of p is thus extended, the expression (3.20) becomes just the series for $J_n(x)$ also for the present case of n, a negative integer. (Proof for the second part, for the case $n < 0$, can also be given indirectly using the proof for the first part for the case $n \geq 0$. Please see Exercise 3.9).

Solved problem 3.4 Establish the recurrence relation

$$2J_n' = J_{n-1} - J_{n+1} \tag{i}$$

for integral n using the generating function.

Solution: We start from

$$\exp\left[\frac{x}{2}\left(t - \frac{1}{t}\right)\right] = \sum_{n=-\infty}^{\infty} t^n J_n(x) \tag{ii}$$

and differentiate both sides with respect to x to obtain

$$\frac{1}{2}\left(t - \frac{1}{t}\right) \exp\left[\frac{x}{2}\left(t - \frac{1}{t}\right)\right] = \sum_{n=-\infty}^{\infty} t^n J_n'(x). \tag{iii}$$

On the left-hand side, we can re-express $\exp\{\frac{x}{2}(t - \frac{1}{t})\}$ using (ii). We then have

$$\frac{1}{2} \sum_{n=-\infty}^{\infty} t^{n+1} J_n - \frac{1}{2} \sum_{n=-\infty}^{\infty} t^{n-1} J_n = \sum_{n=-\infty}^{\infty} t^n J_n'(x). \tag{iv}$$

A comparison of the coefficient of t^n on both sides then gives the desired result (i).

(Note that unlike the derivation in Section 3.3, the derivation presented above is valid only for the *integral order* Bessel functions ($\nu = n$). The other recurrence relation can likewise be obtained from (ii) by the differentiation of (ii) with respect to t (see Exercise 3.8)).

3.6 Real Integral Representation for $J_n(x)$

As a simple application of the generating function for $J_n(x)$, we shall derive the following integral representation for the Bessel functions of integer orders.

$$J_n(x) = \frac{1}{\pi}\int_0^{\pi} \cos(n\theta - x\sin\theta)d\theta \qquad n = 0, 1, 2, \ldots\ldots \tag{3.21}$$

(Such an integral representation should, however, be distinguished from the contour integral representations we shall give later which are valid for any order. The integrals involved in (3.21) are by contrast, ordinary real variable definite integrals.)

To see this, we note that the relation

$$g(x, t) = \sum_{n=-\infty}^{\infty} t^n J_n(x) \tag{3.22}$$

is an identity valid for any t (i.e., real or complex). In particular, we may set $t = e^{i\theta}$, where θ is a real variable. Then, we have,

$$t - \frac{1}{t} = e^{i\theta} - e^{-i\theta} = 2i\sin\theta$$

so that

$$g(x, t) = \exp\left[\frac{x}{2}\left(t - \frac{1}{t}\right)\right] = \exp(ix\sin\theta) = \sum_{m=-\infty}^{\infty} e^{im\theta} J_m(x). \tag{3.23}$$

We multiply both sides by $e^{-in\theta}$ and integrate from $-\pi$ to π. We use,

$$\int_{-\pi}^{\pi} e^{im\theta} e^{-in\theta} d\theta = 2\pi\delta_{mn} \tag{3.24}$$

to obtain

$$J_n(x) = \frac{1}{2\pi}\int_{-\pi}^{\pi} \exp[-ni\theta + ix\sin\theta]d\theta. \tag{3.25}$$

We now take the real part of both sides. Here, we recall that $J_n(x)$ is real. We then obtain

$$J_n(x) = \frac{1}{2\pi}\int_{-\pi}^{\pi} \cos(n\theta - x\sin\theta)d\theta. \tag{3.26}$$

Since the integrand is an even function of θ, we may replace (3.26) by

$$J_n(x) = \frac{1}{\pi}\int_{0}^{\pi} \cos(n\theta - x\sin\theta)d\theta \tag{3.27}$$

thus obtaining the integral representation of (3.21).

In passing, we note that taking the real part of (3.23) leads us to

$$\cos(x\sin\theta) = \sum_{m=-\infty}^{\infty} \cos m\theta\, J_m(x). \tag{3.28}$$

Using $J_{-m}(x) = (-)^m J_m(x)$ and $\cos(-m\theta) = \cos m\theta$, we see that on the right-hand of (3.28) the terms with m odd cancel out when paired with $-m$ and terms with m even add up. This leads us to

$$\cos(x\sin\theta) = J_0(x) + 2\sum_{p=1}^{\infty} \cos(2p\theta)J_{2p}(x). \tag{3.29}$$

In a similar manner, imaginary part of (3.23) leads us ultimately to

$$\sin(x\sin\theta) = 2\sum_{q=1}^{\infty} J_{2q-1}(x)\sin[(2q-1)\theta]. \tag{3.30}$$

Solved problem 3.5

1. Show that $|J_n(x)|$ is bounded for any real x by 1.
2. Verify that the integral representation is consistent with $J_n(x) = (-)^n J_n(x)$.

Solution:

1. $$J_n(x) = \frac{1}{\pi}\int_{0}^{\pi} \cos(n\theta - x\sin\theta)d\theta$$

   We note the fact that the modulus of the integrand $\cos(n\theta - x\sin\theta)$ is bounded by 1. Thus,

   $$|J_n(x)| \le \frac{1}{\pi}\cdot 1 \cdot \int_{0}^{\pi} d\theta = 1.$$

   (In fact, $J_0(x)$ reaches this upper bound only at $x = 0$, and it turns out that $J_n(x)$ with $n \neq 0$ never reaches it. Can you show this from the above discussion?)

2. $$J_{-n}(x) = \frac{1}{\pi}\int_0^{\pi} \cos(-n\theta - x\sin\theta)d\theta$$
$$= \frac{1}{\pi}\int_0^{\pi} \cos(n\theta + x\sin\theta)d\theta$$
$$= J_n(-x)$$
$$= (-)^n J_n(x).$$

3.7 The Neumann and Hankel Functions

As we have seen in Section 1.6, $J_\nu(x)$ and $J_{-\nu}(x)$ form a set of linearly independent solutions to the Bessel equation when ν is not an integer. Any linear combination of these two is, of course, also solution to the Bessel equation. Among these linear combinations there are certain linear combinations (called the Neumann function, the Hankel functions) that have simpler asymptotic properties (and associated physical interpretations) and hence occur frequently. Functions similar to these can be defined also for an integral ν and thus provide a set of linearly independent solutions valid even for an integral ν. In this section we shall discuss these in some detail.

We define Neumann function of order ν for a nonintegral ν by

$$N_\nu(x) = \frac{\cos\nu\pi\, J_\nu(x) - J_{-\nu}(x)}{\sin\nu\pi}. \tag{3.31}$$

This is clearly a linear combination of J_ν and $J_{-\nu}$ and hence a solution to the Bessel equation.

We note that the above definition is however inapplicable when $\nu = n$, an integer. We can however make use of the above definition for $N_\nu(x)$ to construct, by a limiting process, the Neumann functions N_n of any integral order n. $N_n(x)$ so constructed, as we shall see, constitutes a solution to the Bessel equation of order n and is linearly independent of $J_n(x)$. We thus define

$$N_n(x) = \lim_{\nu\to n} N_\nu(x) = \lim_{\nu\to n} \frac{\cos\nu\pi\, J_\nu(x) - J_{-\nu}(x)}{\sin\nu\pi}. \tag{3.32}$$

At $\nu = n$, both the numerator and the denominator vanish (recall: $J_{-n} = (-)^n J_n$). We therefore evaluate N_n by l'Hôpital's rule.

Thus,

$$N_n(x) = \left.\frac{\cos\nu\pi \frac{\partial J_\nu}{\partial\nu} - \frac{\partial}{\partial\nu}J_{-\nu}}{\pi\cos\nu\pi}\right|_{\nu=n}$$
$$= \frac{1}{\pi}\left[\frac{\partial J_\nu}{\partial\nu} - (-)^n \frac{\partial J_{-\nu}}{\partial\nu}\right]_{\nu=n}. \tag{3.33}$$

We shall now show that $N_n(x)$ defined above is, in fact, a solution to the Bessel's equation. To see this, consider the Bessel's equation for order ν:

$$J_\nu'' + \frac{1}{x}J_\nu' + \left(1 - \frac{\nu^2}{x^2}\right)J_\nu = 0$$

We differentiate this equation with respect to ν, and recall that differentiations with respect to ν and x commute. Thus,

$$\left(\frac{\partial J_\nu}{\partial \nu}\right)'' + \frac{1}{x}\left(\frac{\partial J_\nu}{\partial \nu}\right)' + \left(1 - \frac{\nu^2}{x^2}\right)\left(\frac{\partial J_\nu}{\partial \nu}\right) - \frac{2\nu}{x^2}J_\nu = 0. \tag{i}$$

In a similar manner, starting form the differential equation satisfied by $J_{-\nu}(x)$, we obtain

$$\left(\frac{\partial J_{-\nu}}{\partial \nu}\right)'' + \frac{1}{x}\left(\frac{\partial J_{-\nu}}{\partial \nu}\right)' + \left(1 - \frac{\nu^2}{x^2}\right)\left(\frac{\partial J_{-\nu}}{\partial \nu}\right) - \frac{2\nu}{x^2}J_{-\nu} = 0. \tag{ii}$$

We now evaluate [(i) $-$ $(-)^n$(ii)] and take the limit as $\nu \to n$. In this limit, the last terms on the left-hand sides of (i) and (ii) when added together, viz., $\{-\frac{2\nu}{x^2}J_\nu + (-)^n \frac{2\nu}{x^2}J_{-\nu}\}$, vanish because $J_n = (-)^n J_{-n}$. The remaining terms combine to yield

$$N_n'' + \frac{1}{x}N_n' + \left(1 - \frac{n^2}{x^2}\right)N_n = 0, \tag{3.34}$$

proving that N_n is indeed a solution to the Bessel's equation.

Hankel functions

Often, certain linear combinations of the Bessel and the Neumann functions, called the Hankel functions, naturally enter into physics. There are two kinds of Hankel functions. They are defined as

$$H_\nu^{(1)}(x) = J_\nu(x) + iN_\nu(x), \tag{3.35}$$

$$H_\nu^{(2)}(x) = J_\nu(x) - iN_\nu(x). \tag{3.36}$$

We note that for a real x, the two kinds of Hankel functions are complex conjugates of each other: $H_\nu^{(2)}(x) = H_\nu^{(1)*}(x)$. In view of the fact that $H^{(1)}$ and $H^{(2)}$ are *different* linear combinations which are not proportional to each other, they constitute a set of two linearly independent solutions of the Bessel equation. We also note that N_ν is a linear combination of the Bessel functions J_ν and $J_{-\nu}$, each of which satisfies the recurrence relations. It is easy to show that N_ν also satisfies the two recurrence relations of the same form as J_ν. This is true even when $\nu = n$, an integer. The proof is left as an exercise (see Exercise 3.13). As the recurrence relations for the Bessel functions are *linear* in them, it follows that any linear combination of the Bessel function J_ν and N_ν will also satisfy the two recurrence relations of the same form. Thus, the Bessel, the Neumann and the Hankel functions of both types satisfy the recurrence relations given below.

$$F_{\nu-1} = F'_\nu + \frac{\nu}{x} F_\nu \tag{3.37}$$

$$F_{\nu+1} = -F'_\nu + \frac{\nu}{x} F_\nu \tag{3.38}$$

$$F = J, N, H^{(1)}, H^{(2)} \tag{3.39}$$

3.8 Spherical Bessel Functions

We have so far considered the Bessel functions, also called the *cylindrical* Bessel functions. As pointed out in the introduction, these naturally arise in the solution to the radial part of the equation when the Laplace's equation or the Helmholtz equation is separated in *cylindrical* coordinates. On the other hand, when the Helmholtz equation is separated in spherical polar coordinates, the radial part of the solution obeys an ordinary second order differential equation called the spherical Bessel equation. This reads

$$x^2 y'' + 2xy' + [x^2 - n(n+1)]y(x) = 0. \tag{3.40}$$

This equation evidently differs from the Bessel's equation, but can be cast in the form of the Bessel's equation by a change of variables (as we shall see presently). As a result, the solutions to this equation are related to the (cylindrical) Bessel functions and are called spherical Bessel functions.

Conversion to the Bessel equation

Equation (3.40) can be converted to the Bessel equation by the substitution of the *dependent* variable y to u given by

$$u = y\sqrt{x}, \quad \text{i.e.,} \quad y = u/\sqrt{x}. \tag{3.41}$$

Then, we have,

$$y' = u'x^{-\frac{1}{2}} - \frac{1}{2}x^{-\frac{3}{2}}u, \tag{3.42}$$

$$y'' = u''x^{-\frac{1}{2}} - x^{-\frac{3}{2}}u' + \frac{3}{4}x^{-\frac{5}{2}}u. \tag{3.43}$$

Substituting these in (3.40) and cancelling two of the terms and rearranging, one finds

$$x^2u'' + xu' + \left[x^2 - \left(n + \frac{1}{2}\right)^2\right]u = 0. \tag{3.44}$$

(3.44) is just the Bessel's equation of which one solution is the Bessel function of order $\left(n+\frac{1}{2}\right)$. In other words, a solution of (3.44) can be written as

$$u(x) = J_{n+\frac{1}{2}}(x), \tag{3.45}$$

and thus a solution of (3.40) can be written as

$$y = (\text{const})\, x^{-\frac{1}{2}} J_{n+\frac{1}{2}}(x). \tag{3.46}$$

It is conventional to define the normalisation of (3.46) such that

$$y = j_n(x) = \sqrt{\frac{\pi}{2x}}\, J_{n+\frac{1}{2}}(x). \tag{3.47}$$

Then $j_n(x)$ is called the *spherical Bessel function.* (The normalisation of $j_n(x)$ has been chosen so that the asymptotic behaviour of $j_n(x)$ has a simple form. See Section 3.9). There is, of course, another linearly independent solution to (3.40) and we will come to it soon.

$j_n(x)$ *as series in* x

Series for $j_n(x)$ can be obtained by substituting the series for $J_{n+\frac{1}{2}}(x)$ in (3.47). We then have

$$j_n(x) = \sqrt{\frac{\pi}{2x}} \sum_{r=0}^{\infty} \frac{(-)^r}{r!\left(n + \frac{1}{2} + r\right)!} \left(\frac{x}{2}\right)^{n+\frac{1}{2}+2r} \tag{3.48}$$

The denominators now contain factorials of fractions, which is inconvenient (unless n happens to be half odd integer. We shall, on the other hand, be using $j_n(x)$ invariably for an integral n). It can be simplified using the duplication formula for the factorial functions from Chapter 2.

$$2^{2n+2r+1}\left(n + \frac{1}{2} + r\right)!(n + r)! = (2n + 2r + 1)!\sqrt{\pi}$$

Then, after a rearrangement of factors, the series for $j_n(x)$ reads:

$$j_n(x) = 2^n x^n \sum_{r=0}^{\infty} \frac{(-)^r (r + n)!}{r!(2r + 2n + 1)!} x^{2r} \tag{3.49}$$

Spherical Neumann and Hankel functions

$j_n(x)$ defined above constitutes *one solution* to the spherical Bessel equation. There is, of course, a second linearly independent solution. As in the case of the cylindrical Bessel equation, we can define the spherical Neumann and the spherical Hankel functions. These are defined in a manner exactly analogous to the cylindrical Bessel functions. In other words, one has:

$$n_n(x) \equiv \sqrt{\frac{\pi}{2x}} N_{n+\frac{1}{2}}(x) = (-)^{n+1} \sqrt{\frac{\pi}{2x}}\, J_{-n-\frac{1}{2}}(x) \tag{3.50}$$

$$h_n^{(1)}(x) \equiv \sqrt{\frac{\pi}{2x}} H^{(1)}_{n+\frac{1}{2}}(x) = j_n(x) + i n_n(x) \tag{3.51}$$

$$h_n^{(2)}(x) \equiv \sqrt{\frac{\pi}{2x}} H^{(2)}_{n+\frac{1}{2}}(x) = j_n(x) - i n_n(x) \tag{3.52}$$

We can obtain the series for $n_n(x)$, for integral n, in a manner analogous to the one in which we obtained series of (3.49) for $j_n(x)$. We start with the definition (3.50) and the series for $J_{-n-\frac{1}{2}}(x)$. We again employ the duplication formula for simlification and make certain rearrangements. The result reads

$$n_n(x) = \frac{(-)^{n+1}}{2^n x^{n+1}} \sum_{r=0}^{\infty} \frac{(-)^r (r-n)!}{r!(2r-2n)!} x^{2r}. \tag{3.53}$$

Spherical Bessel/Neumann/Hankel functions of zeroth order

We specifically consider these solutions of *order zero*, first because these are easy to recognise in terms of known trigonometric functions and second because these allow us to determine functions of higher (integral) order via a recurrence relation we shall derive.

Consider $j_0(x)$ (see (3.49): set $n = 0$)

$$j_0(x) = \sum_{r=0}^{\infty} \frac{(-)^r}{(2r+1)!} x^{2r} = \frac{\sin x}{x}. \tag{3.54}$$

In a similar manner, (3.54) with $n = 0$ yields

$$n_0(x) = -\frac{1}{x} \sum_{r=0}^{\infty} \frac{(-)^r}{(2r)!} x^{2r} = -\frac{\cos x}{x}. \tag{3.55}$$

Further,

$$h_0^{(1)}(x) = j_0(x) + i n_0(x) = \frac{\sin x - i \cos x}{x} = -i \frac{e^{ix}}{x}, \tag{3.56}$$

and similarly,

$$h_0^{(2)}(x) = i\frac{e^{-ix}}{x}. \tag{3.57}$$

For a real x, the two Hankel functions are related by

$$h_0^{(2)}(x) = h_0^{(1)}(x)^*.$$

(Note that the simple form of results (3.54–3.57) has resulted from the normalisation factor $\sqrt{\frac{\pi}{2}}$ in the definitions (3.47) and (3.50).)

Solved problem 3.6 Verify directly from the definition $j_0'(x) = -j_1(x)$. Show that

$$j_1(x) = -\frac{\cos x}{x} + \frac{\sin x}{x^2}.$$

Solution:

$$j_0(x) = \sum_{r=0}^{\infty} \frac{(-)^r}{(2r+1)!} x^{2r}$$

$$j_0'(x) = \sum_{r=1}^{\infty} \frac{(-)^r 2r}{(2r+1)!} x^{2r-1}$$

$$= \sum_{r=0}^{\infty} \frac{(-)^r(-)(2r+2)}{(2r+3)!} x^{2r+1}$$

$$= -2x \sum_{r=0}^{\infty} \frac{(-)^r (r+1)!}{(2r+3)!r!} x^{2r}$$

$$= -j_1(x)$$

by comparison with (3.49). Using the above result,

$$j_1(x) = -\frac{d}{dx}(j_0(x)) = -\frac{d}{dx}\left(\frac{\sin x}{x}\right)$$

$$= -\frac{\cos x}{x} + \frac{\sin x}{x^2}.$$

Solved problem 3.7 Verify from the differential equation (3.40) for $n = 0$ that $f = xy$ satisfies $f'' + f = 0$. Hence write down the two linearly independent solutions to it. How do you identify the one which refers to $j_0(x)$?

Solution: For $n = 0$, (3.40) reads

$$x^2y'' + 2xy' + x^2y = 0.$$

Hence, this equation leads us to

$$f'' = xy'' + 2y' = -xy = -f.$$

Hence $f = A\sin x + B\cos x$.

$$y = \frac{A\sin x + B\cos x}{x}.$$

Now, $j_0(x)$ is a solution of the equation (3.40) that is finite at $x = 0$. This requires that $B = 0$ in the above. This leads us to

$$j_0(x) \propto \frac{A\sin x}{x}.$$

We can fix the normalisation of y by noting that $j_0(0) = 1$. This fixes A, viz., $A = 1$.

Recurrence relations for $j_n(x)$ and other spherical Bessel functions

The spherical Bessel functions $j_n(x)'s$, being related to the cylindrical ones, also satisfy recurrence relations of similar kinds. As would be expected, these follow simply from those for J_ν with $\nu = n + \frac{1}{2}$. We recall the recurrence relations for J_ν:

$$\frac{d}{dx}(x^\nu J_\nu) = x^\nu J_{\nu-1} \tag{I}$$

$$\frac{d}{dx}(x^{-\nu} J_\nu) = -x^{-\nu} J_{\nu+1} \tag{II}$$

We set $\nu = n + \frac{1}{2}$ and use $J_{n+\frac{1}{2}}(x) = \sqrt{\frac{2x}{\pi}}\, j_n(x)$ in (I) to obtain

$$\frac{d}{dx}\left(x^{n+\frac{1}{2}}\sqrt{\frac{2x}{\pi}}\, j_n(x)\right) = x^{n+\frac{1}{2}}\sqrt{\frac{2x}{\pi}}\, j_{n-1}(x).$$

On simplification, we find

$$\frac{d}{dx}\left(x^{n+1}\, j_n(x)\right) = x^{n+1}\, j_{n-1}(x). \tag{I'}$$

Similarly (II) yields,

$$\frac{d}{dx}(x^{-n} j_n(x)) = -x^{-n} j_{n+1}(x). \tag{II'}$$

These are the two recurrence relations for $j_n(x)$.

As has been established earlier, $N_\nu(x)$ also satisfy the two recurrence relations of the same form as (I) and (II). Hence recurrence relations of the same form as (I′) and (II′) also hold for $n_m(x)$. Consequently, the same conclusion applies to both of the spherical Hankel functions $h_n{}^{(1,2)}(x)$. To summarise, the set of the spherical Bessel functions satisfy the recurrence relations of the form

$$\frac{d}{dx}\left(x^{n+1} f_n(x)\right) = x^{n+1} f_{n-1}(x),$$

$$\frac{d}{dx}\left(x^{-n} f_n(x)\right) = -x^{-n} f_{n+1}(x),$$

where f_n stands for *any one* of j_n, n_n, $h_n^{(1)}$ and $h_n^{(2)}$.

Solved problem 3.8 Establish the following relation from the recurrence relations for $j_n(x)$.

$$j_n(x) = (-x)^n \left(\frac{d}{xdx}\right)^n j_0(x)$$

Solution: From (II′) and with the replacement $n \to n-1$, we have

$$j_n(x) = -x^{n-1} \frac{d}{dx}(x^{-n+1} j_{n-1}(x))$$

$$= (-x)^n \left(\frac{d}{xdx}\right) ((-x)^{-n+1} j_{n-1}(x)).$$

We rewrite it as

$$[(-x)^{-n} j_n(x)] = \left(\frac{d}{xdx}\right) [(-x)^{-(n-1)} j_{(n-1)}(x)].$$

We recognise that the square brackets on both sides are *identical in form* except the replacement $n \to (n-1)$. We thus immediately recognise

$$[(-x)^{-n} j_n(x)] = \left(\frac{d}{xdx}\right)\left(\frac{d}{xdx}\right) [(-x)^{-(n-2)} j_{(n-2)}(x)].$$

Repeating the above procedure $(n-2)$ more times we obtain the result.

Comment: We note that we can therefore write:

$$j_n(x) = (-x)^n \left(\frac{d}{xdx}\right)^n \left[\frac{\sin x}{x}\right]$$

This relation then implies that *all* spherical Bessel functions of *integral* orders can be written entirely in terms of trigonometric functions and powers of $\frac{1}{x}$. In fact, it is not difficult to see that the general structure of $j_n(x)$ will be of the form,

$$j_{2n}(x) = \frac{\sin x}{x} p_1\left(\frac{1}{x^2}\right) + \cos x p_2\left(\frac{1}{x^2}\right)$$

and

$$j_{2n+1}(x) = \frac{\sin x}{x^2} p_3\left(\frac{1}{x^2}\right) + \frac{\cos x}{x} p_4\left(\frac{1}{x^2}\right),$$

where p_1, p_2, p_3 and p_4 are polynomials of $\left(\frac{1}{x^2}\right)$ of degree n. (See Exercise 3.16)

3.9 Bessel and Spherical Bessel Functions: Asymptotic and Small x Behaviour

As mentioned in the introduction to this chapter, the Bessel functions occur in solutions to a large class of physics problems with a cylindrical or a spherical symmetry. In these problems, the general solution to an appropriate Laplace equation or Helmholtz wave equation is written as a linear combination of solutions, each containing appropriate Bessel/Neumann (or Hankel) functions as its radial part. There are several reasons why we would like to know the behaviour of the Bessel type functions near $x \approx 0$ and as $|x| \to \infty$, i.e., asymptotically.

1. The nature of the specific linear combination of Bessel and Neumann functions that can appear in (the radial part of) a solution in a given region of space depends crucially on whether $\rho = 0$ (or $r = 0$) is included in the region. It will turn out that the Neumann (spherical Neumann) function do not occur in a physically meaningful solutions if the region in question includes $\rho = 0$ (or $r = 0$). This arises from the requirement that the physically observable quantities (e.g., amplitude for the vibration of membranes, electrostatic fields, amplitude for a propagating wave, $|\psi|^2$ for a quantum mechanical problem, etc.) must be finite. We shall see that the Neumann functions $N_\nu(x)$ and $n_n(x)$ do not satisfy this property at $x = 0$.
2. While discussing the solution to the wave equation for an outgoing spherical wave, the behaviour of spherical Hankel functions $h_l^{(1,2)}(x)$ as $x \to \infty$ becomes relevant. We shall see that asymptotically $h_l^{(1)}(x)$ represent outgoing waves while $h_l^{(2)}(x)$ represent incoming waves as these behave as $\frac{e^{\pm ix}}{x}$ respectively for large x.

3. While evaluating contribution from various multipoles in the radiation problem in quantum mechanics (in atomic and nuclear physics), the behaviour of spherical Bessel and Hankel functions near $x \approx 0$ and for $x \to \infty$ is needed.

Behaviour of Bessel-type functions near $x = 0$

The behaviour of the Bessel and the spherical Bessel functions near $x = 0$ can simply be read off from the series definition of each of them. The term with lowest power of x gives the leading behaviour.

Thus from (3.7) and (3.13), we read off

$$J_\nu(x) = \frac{x^\nu}{2^\nu \nu!} + O(x^{\nu+1}) \quad \nu \neq -n, \text{ a negative integer}$$

and,

$$J_n(x) = (-1)^n J_n(x) = (-1)^n \frac{x^n}{2^n n!} + O(x^{n+1}).$$

We thus note from the above that

1. for $\nu \geq 0$, $J_0(x)$ alone does not vanish for $x = 0$;
2. for $\nu < 0$, $J_\nu(x)$ blows up near $x = 0$ like $\frac{1}{x^{|\nu|}}$ except when ν is a negative integer; in the latter case $J_{-n}(x) = (-)^n J_n(x)$ vanishes for $x = 0$.

The behaviour of cylindrical Neumann function $N_\nu(x)$ for $(\nu \neq n)$ near $x \sim 0$ can be read off from its definition. For $\nu \neq n$, $J_{-|\nu|}(x)$ term in its definition

$$N_\nu(x) = \frac{\cos \nu\pi \, J_\nu(x) - J_{-\nu(x)}}{\sin \nu\pi}$$

dominates. Thus for $\nu > 0$,

$$N_\nu(x) \simeq -\frac{1}{\sin \nu\pi} \frac{x^{-\nu}}{2^{-\nu}(-\nu)!} = \frac{(\nu-1)!}{\pi} \left(\frac{2}{x}\right)^\nu,$$

where we have used the relation between $(-\nu)!$ and $\nu!$ The above relation also holds for $\nu = n$, a positive integer. For $\nu < 0$, we find,

$$N_\nu(x) \simeq -\frac{\cos \nu\pi (-\nu-1)!}{\pi} \left(\frac{x}{2}\right)^\nu.$$

As for Neumann function of integral order, we obtain from (3.33)

$$N_n(x) = \frac{1}{\pi} \left[\frac{\partial J_\nu}{\partial \nu} - (-)^n \frac{\partial J_{-\nu}}{\partial \nu} \right]_{\nu=n}.$$

To obtain the result for $n = 0$, we note that

$$\left.\frac{\partial J_{-\nu}}{\partial \nu}\right|_{\nu=0} = -\left.\frac{\partial J_{-\nu}}{\partial \nu}\right|_{\nu=0}$$

so that,

$$N_0(x) = \frac{2}{\pi}\left[\frac{\partial J_\nu}{\partial \nu}\right]_{\nu=0} = \frac{2}{\pi}(\ln x + \gamma - \ln 2) + O(x^2).$$

In a similar manner, one can read off the behaviour of the spehrical Bessel functions $j_m(x)$ and $n_m(x)$ from equations (3.48) and (3.53). They are:

$$j_m(x) \simeq x^m \frac{2^m m!}{(2m+1)!} + O(x^{m+2})$$

$$n_m(x) \simeq x^m \frac{(-)^{m+1} 2^m \sqrt{\pi}}{x^{m+1}(-m-1/2)!} + O(x^{-m+1})$$

Behaviour of Bessel-type functions as x → ∞

The asymptotic behaviour of Bessel functions can best be evaluated by using the method of steepest descent as applied to their contour integral representations. We have given a brief discussion on contour integral representations in the next section (Section 3.10.2: see (3.60)). Applying the method of steepest descent (see Section 8.3 and Problem 8.15 of *Mathematical Physics: The Basics*), one obtains that as $x \to \infty$,

$$J_\nu(x) \simeq \sqrt{\frac{2}{\pi x}} \cos\,(x - \theta_\nu); \quad \theta_\nu \equiv \left(\nu + \frac{1}{2}\right)\frac{\pi}{2};$$

$$N_\nu(x) \simeq \sqrt{\frac{2}{\pi x}} \sin\,(x - \theta_\nu);$$

$$H_\nu^{(1,2)}(x) \simeq \sqrt{\frac{2}{\pi x}} \exp\,[\pm i(x - \theta_\nu)].$$

(See also Exercise 3.17). In a similar manner, the asymptotic behaviour of the spherical Bessel functions can be read off from the above using their relation to cylindrical ones (see (3.47) and (3.50)–(3.52)).

$$j_m(x) \simeq \frac{1}{x} \sin\left(x - \frac{m\pi}{2}\right)$$

$$n_m(x) \simeq -\frac{1}{x} \cos\left(x - \frac{m\pi}{2}\right)$$

$$h_m^{(1,2)}(x) \simeq (\mp i)\frac{1}{x} \exp\left[\pm i\left(x - \frac{m\pi}{2}\right)\right]$$

See also Exercise 3.18.

3.10 Summary of some Additional Results

In this section we shall summarise some definitions and results that one may occasionally encounter.

3.10.1 Modified Bessel functions

Modified Bessel functions are related to the Bessel functions with imaginary arguments. Thus,

Definition:

$$I_\nu(x) = e^{-\nu\pi i/2} J_\nu(xe^{\pi i/2}) = \sum_{r=0}^{\infty} \frac{1}{r!(\nu + r)!} \left(\frac{x}{2}\right)^{2r+\nu}.$$

We note that the series for $I_\nu(x)$ is one with all positive coefficients and thus for large x, I_ν increases exponentially. As can be guessed from the form, these constitute one of the solutions to the 'Modified Bessel equation', i.e., one obtained by changing the independent variable to ix. It reads,

$$x^2 y'' + xy' - (x^2 + \nu^2)y = 0. \tag{3.58}$$

A second, linearly independent solution to (3.58) can be obtained as

$$K_\nu(x) \equiv \frac{\pi}{2} i^{\nu+1} H_\nu^{(1)}(ix).$$

The properties of I_ν and K_ν can be obtained from those of J_ν and $H_\nu^{(1)}$. It is evident that these will also satisfy analogous recurrence relations.

3.10.2 Contour integral representations for Bessel functions

We have already seen the contour integral representations for Bessel functions of the integral order. These involve a unit circle as the closed contour. From (3.25), we have,

$$\begin{aligned} J_n(x) &= \frac{1}{2\pi} \int_{-\pi}^{\pi} \exp[-ni\theta + ix \sin\theta] d\theta \\ &= \frac{1}{2\pi i} \oint dt t^{-n-1} \exp\left[\frac{x}{2}\left(t - \frac{1}{t}\right)\right]. \end{aligned} \tag{3.59}$$

We cannot expect a generalisation of the above to nonintegral ν,

$$J_\nu(x) = \frac{1}{2\pi i} \oint dt t^{-\nu-1} \exp\left[\frac{x}{2}\left(t - \frac{1}{t}\right)\right]$$

to work because the integrand has now a branch-cut at $t = 0$ on account of the presence of $t^{-\nu-1}$, and the unit circle crosses it. However, there is a modification of the above that works for all (real) ν and of course reduces to the (3.59) for an integral ν. To write it, we first choose the branch-cut for $t^{-\nu-1}$ in the complex t-plane along the negative real axis. Then we deform the contour so that it does not cut the branch-cut. Thus, the contour C starts at $-\infty$ just *below* the branch-cut; circles the origin in the anticlockwise sense and goes back to $-\infty$ *above* the negative real axis. We thus have the result

$$J_\nu(x) = \frac{1}{2\pi i}\oint_C dt t^{-\nu-1} \exp\left[\frac{x}{2}\left(t - \frac{1}{t}\right)\right]. \tag{3.60}$$

It is not difficult to verify that $J_\nu(x)$ above satisfies the Bessel equation by an explicit evaluation. Further contour integral representations can be derived by the substitution $t = 2s/x$. This leads to a contour integral representation for $J_\nu(x)$:

$$J_\nu(x) = \frac{1}{2\pi i}(x/2)^\nu \oint_{C'} ds s^{-\nu-1} \exp\left(s - \frac{x^2}{4s}\right), \tag{3.61}$$

where C' is a contour in the complex s-plane, identical to the previous one, C. Similar contour integral representations can be written down for the two Hankel functions. These have been used to obtain the asymptotic behaviour of these. (See e.g., Problem 8.15 of *Mathematical Physics: The Basics*.)

3.10.3 Zeros of Bessel functions of integral orders

Zeros of Bessel functions enter into physics frequently. They enter through some sort of boundary condition imposed on the solutions of equations (Laplace's, Helmholtz and Schrödinger) which contain Bessel functions. They are required in diffraction problems, electrostatic boundary value problems, the Deuteron problem, waveguides and so on. One of the first places one encounters the zero of a Bessel function is the formula for the resolving power of a telescope having a circular aperture.

We shall briefly enumerate some of the properties of the zeros of Bessel functions.

(a) Zeros of the Bessel function $J_n(z)$ (z is a complex variable) are isolated. From the series definition we know that the series for $J_n(z)$ converges in the entire complex plane; thus $J_n(z)$ is an entire function. Zeros of an entire function are isolated.
(b) Zeros of $J_n(x)$ (except possibly at $x = 0$) are all simple zeros (see Exercise 3.24)
(c) The Bessel function $J_\nu(x)$ has an infinite number of zeros on the real axis.
(d) Between any two zeros (that does not include $x = 0$) of $J_\nu(x)$ there is one and only one zero of $J_{\nu\pm1}$.
(e) Let x_n be the first positive zero of $J_n(x)$ ($n \geq 0$). Then $x_{n+1} > x_n$.
(f) The zeros of $J_n(x)$ for $|x| \gg n$ are nearly evenly spaced with an interval of π (recall the asymptotic behaviour).

The proofs of these are left as exercises.

Exercises

On § 3.2

3.1 Show that the series for $J_\nu(x)$ is absolutely convergent for all x. Hence draw conclusion about its uniform convergence.

3.2 Show that

$$\frac{1-\cos x}{x} = \int_0^{\pi/2} d\theta\, J_1(x\cos\theta).$$

On § 3.3

3.3 Evaluate $\int_0^\infty dx\, J_1(x) J_0(x)$.

3.4 Show that $\int_0^\infty dx\, J_{2n}(x)$, where n is an integer, has the same value for all n.

3.5 Express J_3 in terms of J_0 and J_1 (Note: The derivatives of J_0 and J_1 should not appear in your answer).

3.6 Show from the recurrence relations for $J_\nu(x)$ that between any two zeros (that does not include $x = 0$) of $J_\nu(x)$ there is one and only one zero of $J_{\nu\pm1}$.

On § 3.4

3.7 Show that when the Laplace's equation is solved by the method of separation of variables in the cylindrical coordinates, one ends up with the Bessel equation (modified Bessel equation) for the radial part. Such solutions are relevant in electrostatics.

On § 3.5

3.8 Starting from the generating function $g(x,t) = \exp\left[\frac{x}{2}\left(t - \frac{1}{t}\right)\right]$ and differentiating with respect to t, obtain the recurrence relation involving Bessel functions of three successive orders: $(n-1), n, (n+1)$.

3.9 Show that the generating function $g(x,t) = \exp\left[\frac{x}{2}\left(t - \frac{1}{t}\right)\right]$ satisfies

$$g(x,t) = g(-x,-t) = g\left(-x, \frac{1}{t}\right).$$

Writing $g(x,t) = \sum_{-\infty}^{\infty} t^n A_n(x)$, show that the above relations imply

(i) $(-)^n A_n(-x) = A_n(x)$,
(ii) $A_{-n}(-x) = A_n(x)$.

Show how the last relation can be used to imply that if $A_n(x)$ is proven to be $J_n(x)$ for $n \geq 0$ (as was done in Section 3.5), then that automatically implies that $A_n(x)$ is equal to $J_n(x)$ also for $n < 0$.

On § 3.6

3.10 Consider the relation for $\cos(x\sin\theta)$ in terms of the even-order Bessel functions. On the right-hand side each term behaves as $\frac{1}{\sqrt{x}}$ asymptotically for large x. See Section 3.9. On the left-hand side there are large values of x for which $\cos(x\sin\theta)$ takes its maximum value $+1$. Examine.

3.11 From the integral representation for $J_n(x)$, show that $|J_n(x)| \leq 1$. Show further that for (i) $n \neq 0$, (ii) for $x \neq 0$ and $n = 0$, the upper bound is never reached.

3.12 Show the expansion of the plane wave $e^{i\mathbf{k}.\mathbf{r}}$ in cylindrical coordinates with $\mathbf{k} = k\hat{i}$

$$e^{i\mathbf{k}.\mathbf{r}} = e^{ik\rho\cos\phi} = 2\sum_{i=0}^{\infty} i^n \cos n\phi J_n(k\rho).$$

On § 3.7

3.13 Show that the Bessel functions $J_\nu, N_\nu, H_\nu^{(1)}$ and $H_\nu^{(2)}$ all satisfy the recurrence relations of the same form.

On § 3.8

3.14 Show that when the Helmholtz wave equation is solved by the method of separation of variables in the spherical polar coordinates, the radial equation is the spherical Bessel equation.

3.15 Show that

$$j_0(kR) + j_2(kR) = \frac{3}{4}\int_{-1}^{1} dt(1-t^2)\exp(ikRt).$$

3.16 Prove the result regarding the general form for the spherical Bessel functions given towards end of Section 3.8

On § 3.9

3.17 To derive the asymptotic behaviour (i.e., as $x \to \infty$) of Bessel/Neumann functions, consider the Bessel equation. Substitute $y = x^p f(x)$. Assume that $|f(x)|$ is bounded as $x \to \infty$ (for an appropriate p) and that the coefficients of powers of x can be compared. Hence show that (i) $p = -\frac{1}{2}$, (ii) f satisfies $f'' + f = 0$, (iii) f can be written as $f = A\cos x + B\sin x$. Verify the self-consistency of the procedure.

3.18 From

$$j_n(x) = (-x)^n \left(\frac{d}{xdx}\right)^n \left(\frac{\sin x}{x}\right),$$

verify the asymptotic behaviour of $j_n(x)$ as $x \to \infty$ as

$$j_n(x) \sim \frac{1}{x} \sin\left(x - \frac{n\pi}{2}\right)$$

3.19 Verify the Wronskian formulas:

(a) $J_\nu(x)J_{-\nu+1}(x) + J_{-\nu}(x)J_{\nu-1}(x) = \dfrac{2\sin\nu\pi}{\pi x}$

(b) $J_\nu(x)N_\nu'(x) - J_\nu'(x)N_\nu(x) = \dfrac{2}{\pi x}$

(c) $J_\nu(x)N_{\nu+1}(x) - J_{\nu+1}(x)N_\nu(x) = -\dfrac{2}{\pi x}$

3.20 Derive the Wronskian relation for the spherical Bessel functions given below.

$$j_n(x)n_n'(x) - j_n'(x)n_n(x) = \frac{1}{x^2}$$

Do this by considering the behaviour near $x = 0$ and also by considering the asymptotic behaviour.

3.21 Show that $\lim_{r\to\infty} \left| k\frac{d}{dk} h_n^{(1)}(kr) \right| = 1$.

On § *3.10*

3.22 Show from the definition of $I_\nu(x)$ that it grows exponentially for large x. Do this by establishing a bound $Ae^{(1-\epsilon)x} < I_\nu(x) < Be^x$ for large enough x with ν, ϵ, A and B positive.

3.23 Show that the Bessel function $J_n(x)$, does not have a double zero, except possibly at $x = 0$. [Hint: For a double zero, $J_n(x)$ and $J_n'(x)$ both vanish but not $J_n''(x)$. Use this in the Bessel equation.]

3.24 Show that all the zeros of $J_\nu(x)$, x real, are simple zeros, except possibly at $x = 0$.

3.25 Show that $J_{\nu+1}$ and $J_{\nu-1}$ cannot have common zeros (with the exception of $\nu = 0$ for which $J_{\pm 1}$ are proportional).

3.26 Let x_n be the first positive zero of $J_n(x)$. Show that $x_{n+1} > x_n$.

3.27 Show that there is no x for which all integral-order Bessel functions $J_n(x)$ vanish simultaneously.

3.28 Show that the Bessel function $J_\nu(x)$ has an infinite number of zeros on the real axis. (Hint: Consider the asymptotic behaviour of $J_\nu(x)$ for large x).

References

[1] Watson, G. N., *A Treatise on the Theory of Bessel Functions*, Cambridge University Press, Cambridge (England), New York, 1995.

4 Some Special Functions and Spherical Harmonics

4.1 Introduction

There are only a few of the commonly occuring bound state problems that are solvable in quantum mechanics. Among these are the standard problems of finding the stationary states of (1) the simple harmonic oscillator; (2) the hydrogen atom; (3) generally a class of spherically symmetric potentials. Each of these problems have a wide range of physical applications and/or extensions. Each of these brings, into physics a class of classical orthogonal polynomials. The simple harmonic oscillator problem requires for its solutions the Hermite polynomials; the hydrogen atom problem requires the Laguerre polynomials, and the spherically symmetric potentials generally require the Legendre polynomials and associated functions. We also stress here that these few solvable problems lend themselves to a large number of physical applications[14] . In addition, the Legendre polynomials and spherical harmonics find applications in a wide class of problems analysed in the spherical polar coordinates. They are found widely useful in electrostatics, multipole expansions, multipole radiation fields, etc. All these sets of orthogonal polynomials (and more) have many properties in common. Each constitutes a set of what is called the 'classical orthogonal polynomials' of successively increasing degree. They satisfy the orthogonality over a certain interval and with a certain (nonnegative) weight function and thus constitute an orthogonal basis for expansion of (a class of) functions over the appropriate interval. Each one satisfies a second order linear homogeneous differential equation (and it is through these that they enter into physical applications). Each has a three-term recurrence relation. Each has a generating function. Each can be expressed by a Rodriguez-type formula. It is then not surprising that they can, in fact, be dealt within a single framework of the so-called classical orthogonal polynomials.

In this chapter we shall present this topic in the following sequence.

We shall first discuss a heuristic approach to the general construction of a set of orthogonal polynomials. We shall then present the rigorous formulation of classical orthogonal

[14]The simple harmonic oscillator wave functions are useful in the exact problems of 3-D harmonic oscillator, Landau levels for a charged particle in a uniform magnetic field, the shell model of the nucleus, and as trial wave functions in variational methods in quantum mechanics. The wave functions for the hydrogen atom are useful in hydrogen-like atoms and for construction of a basis of many-body wave functions in multi-electron atomic systems, etc.

polynomials. We shall subsequently discuss special cases of these that commonly occur in Physics, with a particular emphasis on the Hermite and the Legendre polynomials. Subsequently, we shall elaborately deal with the associated Legendre polynomials and the spherical harmonics. We shall also briefly discuss Laguerre polynomials. Finally, we shall add a summary of the results on other classes of special functions called the confluent hypergeometric functions and hypergeometric functions.

4.2 An Introduction to the Concept of Orthogonal Polynomials

Consider a set of real polynomials of x, viz., $\{C_n(x); n = 0, 1, 2, 3, \dots\}$ whose elements are respectively polynomials of degree n and have the following property:

$$\int_a^b dx C_n(x) C_m(x) w(x) = 0; \quad n \neq m, \tag{4.1}$$

where $[a, b]$ is some interval of the real axis (not necessarily finite) and $w(x)$ is a function of x, nonnegative over $[a, b]$ (and is called a weight function).Then this set constitutes a set of orthogonal polynomials over $[a, b]$ with weight $w(x)$.

We can now consider a subset of this set consisting of first N elements (N arbitrary), viz., $\{C_n(x); n = 0, 1, 2, 3, \dots, (N-1)\}$. We now make several observations regarding this set.

(a) The set $\{C_n(x); n = 0, 1, 2, 3, \dots, (N-1)\}$ is a linearly independent set. This follows from the orthogonality property above. For, if

$$\sum_{r=0}^{N-1} a_r C_r(x) = 0,$$

then this implies,

$$\int_a^b dx C_m(x) w(x) \sum_{r=0}^{N-1} a_r C_r(x) = 0,$$

i.e.,

$$a_m \int_a^b dx w(x) [C_m(x)]^2 = 0; \quad m = 0, 1, 2, \dots, (N-1).$$

In view of the fact that the normalisation integral

$$h_m \equiv \int_a^b dx w(x) [C_m(x)]^2$$

which is an integral of a nonnegative function[15] over $[a, b]$ and hence cannot vanish, we have $a_m = 0$, $m = 0, 1, 2, \ldots, (N-1)$.

(b) The set $\{C_n(x); n = 0, 1, 2, 3, \ldots, (N-1)\}$ therefore serves as an orthogonal basis for expansion of any polynomial in x of degree $\leq (N-1)$.

(c) Equivalently, if we consider an N-dimensional *real* linear vector space V_N of the polynomials in x of degree $(N-1)$ or less, the above set $\{C_n(x); n = 0, 1, 2, 3, \ldots, (N-1)\}$ serves as basis for V_N. This statement is valid for any positive integer N.

(d) The orthogonality of the polynomials then implies that $C_n(x)$ is orthogonal to any polynomial of degree $(n-1)$ or less. In other words,

$$\int_a^b C_n(x)x^p w(x) = 0; \quad p = 0, 1, 2, \ldots, (n-1). \tag{4.2}$$

(e) Let a set of polynomials $\{C_n(x); n = 0, 1, 2, 3, \ldots, (N-1)\}$ be such that each $C_m(x)$ is orthogonal to x^p for $p = 0, 1, \ldots, (m-1)$; then this set is that of orthogonal polynomials in the sense defined above.

To illustrate the basic ideas involved above, we shall consider two illustrations.

Solved problem 4.1 Show that the following polynomials are orthogonal to each other with weight $w(x) = \exp\{-4(x-1)^2\}$ over the interval $(-\infty, \infty)$: (i) $C_1(x) = 4x - 4$ and (2) $C_2(x) = 16x^2 - 32x + 14$. Are they also orthogonal over the interval $[-1, 1]$ with the weight $w(x) = 1$?

Solution: To test the orthogonality of C_1 and C_2 over the interval $(-\infty, \infty)$ with weight $w(x)$, we have to evaluate the integral I below and show that it vanishes.

$$\begin{aligned} I &\equiv \int_{-\infty}^{\infty} C_1(x)C_2(x)w(x)dx \\ &= \int_{-\infty}^{\infty} (4x-4)(16x^2 - 32x + 14)\exp\{-4(x-1)^2\}dx \\ &= \int_{-\infty}^{\infty} (2y)(4y^2 - 2)\exp(-y^2)dy \\ &= 0 \end{aligned}$$

We have made a change of variables $y = 2(x-1)$ and noted that I is an integral of an integrand *odd* in y over the *symmetric* interval $(-\infty, \infty)$.

Further, to check the orthogonality of C_1 and C_2 over the interval $[-1, 1]$ with weight $w(x) = 1$, we consider

$$I' \equiv \int_{-1}^{1} (4x-4)(16x^2 - 32x + 14)dx.$$

[15]Note that $[C_n(x)]^2$ vanishes at at most m points on the real axis.

As the range of integration $[-1, 1]$ is even, we note that we need to keep only the even terms in the integrand. Then we find that $I' = -240 \neq 0$. Thus, we find that C_1 and C_2 are not orthogonal over the interval $[-1, 1]$ with weight $w(x) = 1$.

Solved problem 4.2 Consider a linear vector space of the real polynomials of degree less than or equal to 3. Construct a basis consisting of polynomials $C_n(x)$ of degree n each, with $n = 0, 1, 2$ and 3, and having *definite parity*[16] and orthogonality property

$$\int_{-1}^{1} C_n(x)C_m(x)dx = 0; \quad n \neq m.$$

Show that the results one generates are proportional to the Legendre polynomials of Section 4.6.

Solution: We note that $C_0(x)$ is a constant, and we can choose it to be 1. $C_1(x)$ is a linear function of x and moreover has a definite parity and hence it is proportional to x. We can chose it to be x. Further, $C_2(x)$ is a quadratic polynomial in x and it cannot have a definite parity unless it is even in x. Hence, we let $C_2(x) = ax^2 + b$. Now, $C_2(x)$ is automatically orthogonal to $C_1(x)$ over a symmetric interval $[-1, 1]$ as they have even and odd parities respectively. Also,

$$\int_{-1}^{1} C_2(x)C_0(x)dx = \int_{-1}^{1} (ax^2 + b)dx = \frac{2a}{3} + 2b.$$

This vanishes if $b = -\frac{a}{3}$. Thus, we can choose $a = 3, b = -1$. We thus obtain $C_2(x) = 3x^2 - 1$. A similar logic leads us to the form of $C_3(x) = cx^3 + dx$. This $C_3(x)$ is orthogonal to $C_0(x)$ and $C_2(x)$ on account of their respective parities. We only need to further ensure that

$$\int_{-1}^{1} C_3(x)C_1(x)dx = \int_{-1}^{1} (cx^3 + dx)xdx = \frac{2c}{5} + \frac{2d}{3}$$

vanishes. We could, thus, choose $C_3(x) = 5x^3 - 3x$.

A comparison with the Legendre polynomials $P_n(x)$; $n = 0, 1, 2, 3$ (see section 4.6) shows that

$$C_0 = P_0; \quad C_1 = P_1; \quad C_2 = 2P_2; \quad \text{and} \quad C_3 = 2P_3.$$

Independent sets of orthogonal polynomials

Consider a set of orthogonal polynomials in the interval (a, b) with a weight function $w(x)$:

$$\{C_n(x); \quad n = 0, 1, 2, \ldots\}$$

[16] A function $f(x)$ such that $f(-x) = \pm f(x)$ is said to have *definite* (even/odd) parity.

Consider a linear change of variables: $x = cy + d(c \neq 0)$. We re-express $C_n(x) = C_n(cy + d) = \tilde{C}_n(y)$. Then,

(a) $\tilde{C}_n(y)$ itself is a polynomial of degree n.
(b) $\tilde{C}_n(y)$ and $\tilde{C}_m(y)$ are orthogonal to each other $(m \neq n)$ with weight function $w'(cy+d)$ and over an interval bounded by $y = \frac{a-d}{c}$, and $y = \frac{b-d}{c}$.

Thus, these also constitute a new set of orthogonal polynomials over a (possibly) different interval and with a (possibly) different weight function that is derivable from the former. We shall not consider these as 'independent' sets of orthogonal polynomials. In view of the above possibility, we can always choose to map the interval (a, b) by a linear transformations as follows:

(a) With a, b finite, map to $(-1, 1)$
(b) With only a finite, map to $(0, \infty)$
(c) The remaining case is of course of $(a, b) \equiv (-\infty, \infty)$.

We shall call these the 'standard' intervals.

4.3 A Heuristic Approach to Orthogonal Polynomials

Let us suppose that we want to construct an infinite set of orthogonal polynomials $\{C_n(x);\ n = 0, 1, 2, \ldots\}$, where $C_n(x)$ is a polynomial of nth degree such that they are orthogonal to each other over the interval $[a, b]$ with weight function w:

$$\int_a^b C_n(x)C_m(x)w(x)dx = 0 \quad n \neq m$$

Consider a particular $C_n(x)$. On account of the orthogonality property it must be orthogonal to x^p for $p = 0, 1, 2, \ldots, (n-1)$ [as we have seen in Section 4.2 (see (4.2))]. In other words,

$$\int_a^b [C_n(x)w(x)]x^p dx = 0 \qquad p = 0, 1, 2, \ldots, (n-1). \tag{4.3}$$

To see how such a set of conditions can be realised, consider first the simple case $n = 1$, $p = 0$. We then have

$$\int_a^b [C_1(x)w(x)]dx = 0. \tag{4.4}$$

Now, one simple way to arrange this is to have

$$w(x)C_1(x) = \frac{d}{dx}(X_1) \tag{4.5}$$

and require that

$$X_1(a) = 0 = X_1(b).$$

The latter condition can be arranged, for example, by requiring that

$$X_1(x) \sim (x-a)(x-b).$$

While this may not be the only way to satisfy (4.4), this procedure is amenable to a generalization for all n. For example, consider the two conditions for $n = 2$ with $p = 0, 1$. They are

$$\int_a^b [C_2(x)w(x)]dx = 0,$$

$$\int_a^b [C_2(x)w(x)]x dx = 0. \tag{4.6}$$

In order that the second of these is fulfilled (in a manner *similar* to the previous case), we now require that $w(x)C_2(x)$ is a *second* derivative of a function $X_2(x)$. Moreover, in order that the first condition is satisfied, this function need have an *additional* factor of $(x-a)(x-b)$. We thus write,

$$w(x)C_2(x) = \frac{d^2}{dx^2}(X_2) \tag{4.7}$$

and require that

$$X_2(x) \sim (x-a)^2(x-b)^2.$$

We shall now address to the generalization of this procedure for an arbitrary n. A generalization of the above that allows one to fulfill (4.3) is,

$$w(x)C_n(x) = \frac{d^n}{dx^n}(X_n). \tag{4.8}$$

For then, the vanishing of

$$\int_a^b \frac{d^n}{dx^n}(X_n)\, x^p dx$$

can be expressed entirely in terms of the boundary terms of the type given below.

$$\left[\frac{d^{n-1}}{dx^{n-1}}(X_n)\, x^p - p\frac{d^{n-2}}{dx^{n-2}}(X_n)\, x^{p-1} + \cdots + p!(-)^p \frac{d^{n-p-1}}{dx^{n-p-1}}(X_n)\right]_a^b = 0. \tag{4.9}$$

We must require the above conditions for $p = 0, 1, 2, \dots, (n-1)$. Now, for $p = 0$, we have

$$\left.\frac{d^{n-1}}{dx^{n-1}}(X_n)\right|_a^b = 0.$$

For $p = 1$, we have

$$\left[\frac{d^{n-1}}{dx^{n-1}}(X_n)\,x - \frac{d^{n-2}}{dx^{n-2}}(X_n)\right]_a^b = 0$$

and so on. An general way to arrange these conditions is to require that X_n and each of its first $(n-1)$ derivatives vanish at the end-points of the interval.

$$\left.\frac{d^q}{dx^q}(X_n)\right|_a = 0 = \left.\frac{d^q}{dx^q}(X_n)\right|_b \qquad q = 0, 1, \dots, (n-1)$$

We can have all these derivatives of X_n vanish at both the end points of the interval $[a, b]$ if it has n factors of $(x-a)$ and $(x-b)$ each.

$$X_n(x) \sim (x-a)^n (x-b)^n f_n(x) \equiv s^n f_n(x)$$

The above treatment needs modification if one of the ends of the intervals is $\pm\infty$. Let us suppose, for concreteness, that a is $-\infty$. Then we do not (and cannot) require the presence of a factor $(x-a)^n$ in X_n. We will have to require that the function $f_n(x)$ takes care of the vanishing of X_n and its derivatives up to $(n-1)^{th}$ order at the lower end of the interval $(-\infty)$. In order that this be valid, $f_n(x)$ must vanish at $-\infty$ faster than $(x-b)^{-n}$. In this case, we expect the form of X_n as

$$X_n \sim (x-b)^n f_n(x),$$

with $f_n(x)$, together with its first $(n-1)$ derivatives vanishing fast enough as $x \to -\infty$.

Similarly, in the remaining case when the interval $[a, b]$ is identified with $(-\infty, \infty)$, we expect the form

$$X_n \sim f_n(x)$$

with $f_n(x) \to 0$ fast enough as $x \to \pm\infty$. Thus, in all the three cases we find that

$$w(x)C_n(x) = \frac{d^n}{dx^n}(X_n), \tag{4.10}$$

$$X_n(x) \sim s^n(x) f_n(x), \tag{4.11}$$

with a polynomial s that has the following properties:

(a) It is a quadratic function vanishing at ends of the interval (a, b).
(b) Or, it is a linear function vanishing at the *finite* end $x = a$ of the interval (a, ∞) or $(-\infty, a)$.
(c) Or, it is a constant when the interval is $(-\infty, \infty)$.
(d) In each of the three cases, we can choose the intervals in the 'standard' form (see Section 4.2). Then we can choose $s = (x^2 - 1)$ or x or 1 in the three cases respectively.

We note that the form (4.10) of $C_n(x)$'s does not as yet guarantee that these will be *polynomials*. We can however conclude from the above discussion that if we consider *functions* given by

$$C_n(x) = \frac{1}{w(x)} \frac{d^n}{dx^n} \left(s^n(x) f_n(x)\right)$$

with $s(x)$ and /or $f_n(x)$ vanishing suitably at the end points of an interval (a, b) as spelt out earlier, then,

(a) $\int_a^b dx w(x) C_n(x) x^p = 0 \qquad p = 0, 1, 2, \ldots, (n-1).$
(b) Further, if the $C_n(x)$ are in fact polynomials of nth degree, then the above implies

$$\int_a^b dx w(x) C_n(x) C_p(x) = 0 \qquad p = 0, 1, 2, \ldots, (n-1).$$

4.4 General Formulation of Classical Orthogonal Polynomials

In the previous section we motivated the features we expect of a general formulation of orthogonal polynomials. We shall now give a general formulation of the classical orthogonal polynomials.

4.4.1 Definition of classical orthogonal polynomials over an interval (a, b)

These are defined in terms of a nonnegative real weight function $w(x)$ and a polynomial $s(x)$ of degree 2 or less, with real roots. In the past section, we have already motivated the occurrence of an $s(x)$ with these properties. The nth degree polynomial $C_n(x)$ is given by,

$$C_n(x) = \frac{1}{w(x)} \frac{d^n}{dx^n} \left[ws^n\right], \tag{4.12}$$

where w and s are required to further satisfy

(a) $w(x)$ is finite and infinitely differentiable within (a, b);
(b) $w(a)s(a) = 0 = w(b)s(b)$ whenever a and/or b are finite; and $w(x)s(x) \to 0$ faster than any power of $1/x$ whenever a and/or b are $\pm\infty$.

(c) $C_1(x) = \frac{1}{w(x)}\frac{d}{dx}[ws]$ is linear polynomial in x.
We shall show that such a set of functions $C_n(x)$'s of (4.12) constitute a set of orthogonal polynomials over the interval (a, b) with weight $w(x)$.

We note that the weight functions defined above is always integrable over the associated interval (a, b) in the sense that

$$\int_a^b dx w(x) < \infty$$

as explained below.

(a) If an end-point of the interval, say a, is finite, the condition $w(a)s(a) = 0$ there requires that $w(x)(x - a) \to 0$ as $x \to a$ (from within the interval). This implies a behaviour of $w(x)$ near $x = a$ such as to ensure the convergence of the above integral at the end $x = a$.
(b) For the case where the interval is infinite, $w(x) \to 0$ as $x \to \pm\infty$ faster than any power of $1/x$. Thus, convergence at the infinite end-point(s) is assured.
(c) Further, $w(x)$ is finite in (a, b). These are sufficient to assure integrability.

We further note that the definition of $C_n(x)$ is unaffected by a scaling of $w(x)$ by a positive constant. The orthogonality property is also not affected by the scaling of $s(x)$.

It turns out that there are but a few 'independent' types[17] of solutions to $w(x)$ satisfying the requirements listed above.

To obtain the possible set of weight functions, we consider the equation

$$C_1(x) = \frac{1}{w(x)}\frac{d[w(x)s(x)]}{dx} = ax + b \tag{4.13}$$

in the three cases of the 'standard' intervals. This amounts to solving a first order ODE for $w(x)$:

$$\frac{1}{w(x)}\frac{dw(x)}{dx}s(x) = ax + b - \frac{ds(x)}{dx} = \text{a linear polynomial.} \tag{4.14}$$

1. **Case 1: $(a, b) \equiv (-\infty, \infty)$**
In this case, we still retain the freedom of making a linear transformation that will cast $C_1(x) \equiv \pm 2x$. Thus, we have to solve (note that $s = 1$),

$$\frac{1}{w(x)}\frac{dw(x)}{dx} = \pm 2x.$$

[17] By independent, we mean leading to independent sets of polynomials in the sense of Section 4.2.

This has the solutions

$$w(x) = C \exp(\pm x^2).$$

Nonnegativity of $w(x)$ restricts C to be positive and boundary conditions at $\pm\infty$ require the negative sign in the exponent. Further, as a scaling of $w(x)$ by a positive constant does not affect $C_n(x)$, we find the 'standard' form for $w(x)$ and $C_1(x)$ as:

$$w(x) = \exp(-x^2), \quad C_1(x) = 2x.$$

We can now see by inspection that the C_n's defined by the (4.12), viz.,

$$C_n(x) \equiv \exp(x^2)\frac{d^n}{dx^n}\left[\exp(-x^2)\right]$$

are polynomials indeed (see Exercise 4.2).

2. **Case 2:** $(a, b) \equiv (0, \infty)$
   In this case, we have to solve

$$C_1(x) = \frac{1}{w(x)}\frac{d[w(x)x]}{dx} = ax + b.$$

This leads to (see Exercise 4.1),

$$w(x) = Ce^{ax}x^{b-1}.$$

The boundary condition at $x \to \infty$ requires $a < 0$. Boundary conditions on $w(x)s(x)$ at $x = 0$ require $b > 0$. Positivity of $w(x)$ requires $C > 0$ and using the ability to scale $w(x)$ as well as x, it leads to the standard form:

$$w(x) = e^{-x}x^{\rho}, \quad \rho > -1.$$

In this case also, it is easy to see by inspection that $C_n(x)$ defined by

$$C_n(x) \equiv \exp(x)x^{-\rho}\frac{d^n}{dx^n}\left[\exp(-x)x^{n+\rho}\right]$$

are polynomials indeed (see Exercise 4.2).

3. **Case 3:** $(a, b) \equiv (-1, 1)$
   In the present case, we have to solve

$$C_1(x) = \frac{1}{w(x)}\frac{d[w(x)(x^2 - 1)]}{dx} = ax + b.$$

We can solve this and find (see Exercise 4.1) that the weight function has the form

$$w(x) = C(1 - x)^{\alpha}(x + 1)^{\beta} \quad -1 < x < 1.$$

Boundary conditions on $w(x)s(x)$ at $x = \pm 1$ are satisfied if $\alpha, \beta > -1$. Under these conditions, $w(x)$ is integrable on the interval $[a, b]$. Nonnegative nature of $w(x)$ requires that $C > 0$, and we can choose it to be one. We can again see by inspection that the C_n's defined by the (4.12), viz.,

$$C_n(x) \equiv (1-x)^{-\alpha}(x+1)^{-\beta}\frac{d^n}{dx^n}\left[(1-x)^{\alpha}(x+1)^{\beta}(x^2-1)^n\right]$$

$$= (-)^n(1-x)^{-\alpha}(x+1)^{-\beta}\frac{d^n}{dx^n}\left[(1-x)^{\alpha+n}(x+1)^{\beta+n}\right]$$

are polynomials (see Exercise 4.2).

Finally, we note that we can always scale the polynomials so constructed by a nonzero constant without affecting the degree and the orthogonality properties. We shall therefore define

$$C_n(x) = \frac{1}{N_n}\frac{1}{w(x)}\frac{d^n}{dx^n}\left[ws^n\right], \tag{4.15}$$

where the normalisation constants N_n will later be chosen in each case separately, primarily to suit the convenience of usage.

4.4.2 Three-Term recurrence relations

We shall now establish that the set of orthogonal polynomials must satisfy a recurrence relation involving polynomials of three successive degrees. We express

$$C_n(x) = \sum_{r=0}^{n} a_n^r x^r. \tag{4.16}$$

Now, consider the expression $C_{n+1}(x) - \frac{a_{n+1}^{n+1}}{a_n^n} x C_n(x)$. In this combination, the terms of $O(x^{n+1})$ cancel out and thus it constitutes a polynomial of degree n or less. We can then expand

$$C_{n+1}(x) - \frac{a_{n+1}^{n+1}}{a_n^n} x C_n(x) = \sum_{i=1}^{n} \delta_i C_i(x). \tag{4.17}$$

Now, we note that $C_{n+1}(x)$ is orthogonal to all $C_i(x)$ for $i = 0, 1, \ldots, n$. Further, $xC_n(x)$ is orthogonal to all $C_i(x)$ for $i = 0, 1, \ldots, n-2$. This is so since

$$\int_a^b dx w(x)[xC_n(x)]C_i(x) = \int_a^b dx w(x)C_n(x)[xC_i(x)]$$

$$= 0 \qquad i = 0, 1, \ldots, n-2.$$

The last step follows from the fact that $C_n(x)$ is orthogonal to all polynomials of degree less than n, in particular, $xC_i(x)$ for $i = 0, 1, \ldots, n-2$. Thus, the left-hand side of (4.17) is orthogonal to all $C_i(x)$ with $i = 0, 1, \ldots, n-2$. Therefore, only two terms on the right-hand side of (4.17) can survive, those corresponding to $i = n-1$ and $i = n$. These, in turn, imply what is termed a 'three-term recurrence relation' which has the form,

$$C_{n+1} - \alpha_n x C_n(x) = \beta_n C_n(x) + \gamma_n C_{n-1}(x). \tag{4.18}$$

The value of α_n has already been given in (4.17). The above constitutes a relation between classical orthogonal polynomials of three successive degrees in a given set. (In addition, it turns out that the individual sets of orthogonal polynomials each have another recurrence relation characteristic to each set.)

4.4.3 Differential equations satisfied by the $C_n(x)$

We have emphasised in the beginning of this section that the classical orthogonal polynomials enter into physics primarily through the solutions to differential equations occurring in physics. To make this connection credible, we shall now show that each of the classical orthogonal polynomials in fact satisfies a linear homogeneous second order ODE characteristic to it. To see this, we start from

$$F_n(x) \equiv \frac{1}{w(x)}\frac{d}{dx}\left[ws\frac{d}{dx}C_n(x)\right]$$

and show that it is a multiple of $C_n(x)$ itself. To see this we establish successively that

(i) $F_n(x)$ is a polynomial of degree $\le n$,
(ii) it is orthogonal to $C_i(x)$ for each $i = 0, 1, \ldots, n-1$, and thus,
(iii) it is a polynomial of degree n, if non-vanishing.

To prove these, we first expand out $F_n(x)$:

$$F_n(x) = \left[\frac{1}{w}\frac{dw}{dx}s\right]\frac{d}{dx}C_n(x) + \frac{ds}{dx}\frac{d}{dx}C_n(x) + s\frac{d^2}{dx^2}C_n(x) \tag{4.19}$$

and note that each term in its expansion is a polynomial of degree $\le n$ (note the equation (4.14)). To see (ii), we consider

$$\begin{aligned} I_{np} &= \int_a^b dx\, w(x) F_n(x) C_p(x) \\ &= \int_a^b dx \frac{d}{dx}\left[ws\frac{d}{dx}C_n(x)\right] C_p(x) \end{aligned}$$

$$= \int_a^b dx \left[ws \frac{d}{dx} C_n(x) \right] \frac{d}{dx} C_p(x). \tag{4.20}$$

Above, we have employed an integration by parts and imposed the boundary conditions that $w(x)s(x)$ satisfies. We note that the above expression is symmetric in (n, p) and vanishes when $p > n$ on account of orthogonality between $C_p(x)$ and any polynomial of degree less than p such as $F_n(x)$. Thus it also vanishes when $n > p$, by the symmetry of I_{np}. Hence $F_n(x)$, a polynomial of degree n, orthogonal to all those C_i's of lesser degrees, must be proportional to $C_n(x)$, if non-vanishing (complete the argument by expanding $F_n(x)$ in terms of $C_i(x)$'s). It thus follows that

$$\frac{1}{w(x)} \frac{d}{dx} \left[ws \frac{d}{dx} C_n(x) \right] = \lambda C_n(x), \tag{4.21}$$

where λ is some constant. This constitutes a linear homogeneous second order ODE satisfied by $C_n(x)$.

4.5 Hermite Polynomials

As stated earlier, the Hermite polynomials occur as a special case of the general classical orthogonal polynomials discussed earlier. In the general expression (4.15) for $C_n(x)$,

$$C_n(x) = \frac{1}{wN_n} \frac{d^n}{dx^n}(ws^n); \quad n = 0, 1, 2, \ldots$$

we choose,

$$s = 1; \quad w = \exp(-x^2) \quad (a, b) \equiv (-\infty, \infty); \quad N_n = (-)^n.$$

We then obtain the so-called Hermite polynomials, viz.,

$$H_n(x) = (-)^n \exp(x^2) \frac{d^n}{dx^n}[\exp(-x^2)]. \tag{4.22}$$

A first few Hermite polynomials are

$$H_0(x) = 1; \qquad H_1(x) = 2x;$$

$$H_2(x) = 4x^2 - 2; \quad H_3(x) = 8x^3 - 12x.$$

We note that H_n, which is a polynomial of degree n, is from the definition (4.22) either even or odd depending on whether n is even or odd. Thus, H_n contains either only even or only odd powers of x.

Hermite polynomials are orthogonal to each other with the weight $w(x)$:

$$\int_{-\infty}^{\infty} H_m(x) H_n(x) \exp(-x^2) dx = 0; \quad m \neq n$$

and the specific choice of N_n above amounts to the normalisation integral

$$\int_{-\infty}^{\infty} H_n^2(x)\exp(-x^2)dx = 2^n\sqrt{\pi}n!$$

This is left as an exercise (see Exercise 4.4). The Hermite polynomials satisfy the 3-term recurrence relation (4.18), which as applied to this case reads

$$H_{n+1} = 2xH_n(x) - 2nH_{n-1}(x).$$

We note that the '$\beta_n C_n(x)$' term from (4.18) is missing as β_n vanishes in this case, on account of definite parity of $H_n(x)$'s (see Exercise 4.3). In addition, the Hermite polynomials have another recurrence relation (characteristic only to them):

$$H_n'(x) = 2nH_{n-1}(x). \tag{4.23}$$

(We shall give a derivation of this later.) The Hermite polynomials satisfy the differential equation

$$H_n''(x) - 2xH_n'(x) + 2nH_n(x) = 0. \tag{4.24}$$

One of the important application of the Hermite polynomials appears in quantum mechanical solution to the Schrödinger equation for a simple harmonic oscillator. There, it is found that the wave function $\psi_n(x)$ for the nth excited state is proportional to the Hermite polynomial $H_n(x)$:

$$\psi_n(x) = (const)H_n(\alpha x)\exp(-\alpha^2 x^2),$$

where α is defined in terms of the mass m and angular frequency ω as

$$\alpha = \sqrt{\frac{m\omega}{\hbar}}.$$

4.5.1 The generating function for the Hermite polynomials

We have already come across the idea of a 'generating function' while dealing with the Bessel functions of the integral order. There, we had discussed and become familiar with the advantages of a generating function. We shall now show that the set of Hermite polynomials can also be generated from a generating function and can simultaneously be dealt with with its help.

Consider the function

$$g(x,t) = \exp(2xt - t^2). \tag{4.25}$$

We shall show that $g(x,t)$ generates the Hermite polynomials in the sense that when g(x,t) is expanded in (nonnegative) powers of t, the coefficients yield the successive Hermite polynomials.

$$g(x,t) = \sum_{n=0}^{\infty} \frac{t^n}{n!} H_n(x) \tag{4.26}$$

or equivalently,

$$H_n(x) = \left. \frac{d^n}{dt^n} g(x,t) \right|_{t=0}.$$

To prove this result, we shall expand out $g(x,t)$ in powers of t and show that the coefficient of $\frac{t^n}{n!}$ is $H_n(x)$, as defined by the generalised Rodriguez formula of (4.22). We write

$$g(x,t) = \exp(2xt - t^2) = \exp[-(t-x)^2 + x^2].$$

Now,

$$\left. \frac{d^n}{dt^n} g(x,t) \right|_{t=0} = \exp(x^2) \frac{d^n}{dt^n} \left\{ \exp[-(t-x)^2] \right\} \Big|_{t=0}.$$

We now let $y = t - x$. Then,

$$\frac{d}{dt} f(y) = \frac{d}{dy} f(y) = -\frac{d}{dx} f(y).$$

Or, generally,

$$\frac{d^n}{dt^n} f(y) = (-)^n \frac{d^n}{dx^n} f(y).$$

On the left-hand side, we set $t = 0$ after differentiation.

$$\left[\frac{d^n}{dt^n} f(y) \right] \Big|_{t=0} = (-)^n \left[\frac{d^n}{dx^n} f(y) \right] \Big|_{t=0} = (-)^n \frac{d^n}{dx^n} [f(y; t=0)]$$

$$= (-)^n \frac{d^n}{dx^n} f(-x). \tag{4.27}$$

In the second step, we have brought $t = 0$ inside the differentiation (which is with respect to x) and then have used $f(y, t=0) = f(-x)$. Now, letting $f(y) = \exp(-y^2)$ in (4.27), we obtain

$$\left. \frac{d^n}{dt^n} \exp[-(t-x)^2] \right|_{t=0} = (-)^n \frac{d^n}{dx^n} \exp(-x^2).$$

This, in turn, leads us to

$$\left. \frac{d^n}{dt^n} g(x,t) \right|_{t=0} = \exp(x^2)(-)^n \frac{d^n}{dx^n} \exp(-x^2) = H_n(x).$$

This completes the proof that $g(x, t)$ generates the Hermite polynomials as indicated in (4.26).

To become more familiar with the generating function and to illustrate its usage, we shall derive the recurrence relations mentioned earlier with its help.

Solved problem 4.3 Using the generating function $g(x, t)$, derive the recurrence relation for the Hermite polynomials:

(a) $H'_n(x) = 2nH_{n-1}(x)$

(b) $H_{n+1} = 2xH_n(x) - 2nH_{n-1}(x)$

Solution: We start from the generating function

$$g(x, t) = \exp(2xt - t^2) = \sum_{n=0}^{\infty} \frac{t^n}{n!} H_n(x).$$

(a) We differentiate the relation with respect to x to obtain

$$\frac{\partial}{\partial x} g(x, t) = (2t)\exp(2xt - t^2) = \sum_{n=0}^{\infty} \frac{t^n}{n!} H'_n(x).$$

We write the above equation as

$$2\sum_{n=0}^{\infty} \frac{t^{n+1}}{n!} H_n(x) = 2\sum_{n=1}^{\infty} \frac{t^n}{(n-1)!} H_{n-1}(x) = 2\sum_{n=0}^{\infty} \frac{t^n n}{n!} H_{n-1}(x)$$

$$= \sum_{n=0}^{\infty} \frac{t^n}{n!} H'_n(x).$$

Comparing the coefficients of the power t^n on both sides, we obtain

$$H'_n(x) = 2nH_{n-1}(x).$$

(b) We now differentiate the relation (4.26) with respect to t to obtain

$$\frac{\partial}{\partial t} g(x, t) = 2(x - t)\exp(2xt - t^2) = \sum_{n=0}^{\infty} \frac{nt^{n-1}}{n!} H_n(x).$$

We rearrange the above relation as

$$2x\sum_{n=0}^{\infty} \frac{t^n}{n!} H_n(x) - 2\sum_{n=0}^{\infty} \frac{t^{n+1}}{n!} H_n(x) = \sum_{n=0}^{\infty} \frac{nt^{n-1}}{n!} H_n(x).$$

Rearranging the summation indices,

$$2x\sum_{n=0}^{\infty}\frac{t^n}{n!}H_n(x)-2\sum_{n=0}^{\infty}\frac{t^n n}{n!}H_{n-1}(x)=\sum_{n=0}^{\infty}\frac{t^n}{n!}H_{n+1}(x).$$

A comparison of powers of t^n leads us to the second recurrence relation (b).

Comment: We note that the method outlined above will also be useful in establishing the recurrence relations in the case of the Legendre polynomials and the Laguerre polynomials that we shall deal with in subsequent sections.

4.6 Legendre Polynomials

Legendre polynomials are perhaps the most frequently occurring classical orthogonal polynomials in Physics. This is due mainly to the fact that they occur naturally in the process of obtaining the solution to the angular part of the differential equation involving the Laplacian, ∇^2, such as those we have discussed at the beginning of Chapter 1, viz., the Laplace's equation, Helmholtz wave equation, the Schrödinger equation for a spherically symmetric potential, etc.

As mentioned earlier, the set of Legendre polynomials are understood as a special case of the classical orthogonal polynomials of (4.15) if we set $w = 1$ and $s = x^2 - 1$ and with the interval $[a, b]$ identified as $[-1, 1]$. We thus have the Rodriguez formula,

$$P_n(x)=\frac{1}{2^n n!}\frac{d^n}{dx^n}[(x^2-1)^n]=\frac{(-)^n}{2^n n!}\frac{d^n}{dx^n}[(1-x^2)^n], \tag{4.28}$$

where, we have chosen the normalisation N_n of $P_n(x)$ as given above, for reasons to be explained later. A first few of the Legendre polynomials are:

$$P_0(x)=1;\quad P_1(x)=x;$$

$$P_2(x)=\frac{1}{2}(3x^2-1);\quad P_3(x)=\frac{1}{2}(5x^3-3x).$$

From the definition, it is evident that $P_n^{\prime}s$ are polynomials of degree n and are either even or odd according to whether n is even or odd, i.e.,

$$P_n(-x)=(-)^n P_n(x).$$

The Legendre polynomials are orthogonal in the sense of (4.1), i.e.,

$$\int_{-1}^{1}P_n(x)P_m(x)dx=0;\quad n\neq m$$

and they have been normalised by the convention (see Exercise 4.17),

$$P_n(1) = 1.$$

This, in turn, implies the normalisation integral

$$\int_{-1}^{1} [P_n(x)]^2 dx = \frac{2}{2n+1}$$

(see Exercise 4.19). The Legendre polynomials satisfy the special case of the general 3-term recurrence relation (4.18). In this case, it becomes

$$(n+1)P_{n+1}(x) = (2n+1)xP_n(x) - nP_{n-1}(x). \tag{4.29}$$

As remarked earlier in the case of the Hermite polynomials, the above recurrence relation also lacks the $\beta_n C_n$ term on account of the fact that the Legendre polynomials have a definite parity. They also satisfy the ODE (4.21) as applied to this special case, viz.,

$$(1-x^2)P_n''(x) - 2xP_n'(x) + n(n+1)P_n(x) = 0. \tag{4.30}$$

The Legendre polynomials are the only *polynomial* solutions of the above equation and these exist only when n is a nonnegative integer (see e.g., Exercise 1.25 of Chapter 1 for more details). As in the case of the Hermite polynomials, the Legendre polynomials have an additional recurrence relation characteristic to them. It reads:

$$P_n(x) = P'_{n+1} + P'_{n-1} - 2xP'_n(x). \tag{4.31}$$

This can be shown with the help of the Rodriguez formula, or better yet, the generating function which we will introduce next.

4.6.1 The generating function for the Legendre polynomials

The Legendre polynomials can be generated by a rather simple generating function

$$g(x,t) \equiv (1-2xt+t^2)^{-1/2} = \sum_{n=0}^{\infty} t^n P_n(x); \quad |t| < 1, \tag{4.32}$$

or, in other words,

$$P_n(x) = \frac{1}{n!}\frac{d^n}{dt^n} g(x,t)\bigg|_{t=0}.$$

This form of the generating function is easily proved by expanding it in powers of t and comparing the coefficients of t^n with the *expansion* of the Rodriguez formula. See Problem 4.20 for further details.

Generating function and the coulomb potential

There is a close relation between the expression for the Coulomb potential at $\mathbf{r}$ of a charge located at $\mathbf{r}'$, viz.,

$$\Phi(\mathbf{r}) = C\frac{1}{|\mathbf{r}-\mathbf{r}'|}$$

(where C is a constant). This is not surprising because the above Coulomb potential $\Phi(\mathbf{r})$ is a solution of the Laplace's equation $\nabla^2\Phi = 0$ in the domain that excludes the point $\mathbf{r}'$, whereas the Legendre polynomials are naturally occurring in the solution of the same equation. To see the connection, we rewrite,

$$\begin{aligned}\frac{1}{|\mathbf{r}-\mathbf{r}'|} &= (r^2 + r'^2 - 2rr')^{-1/2} \\ &= \frac{1}{r}\left[1 + \left(\frac{r'}{r}\right)^2 - 2\left(\frac{r'}{r}\right)\cos\theta\right]^{-1/2} \\ &= \sum_{l=0}^{\infty}\frac{r'^l}{r^{l+1}}P_l(\cos\theta).\end{aligned}$$

The last step is possible for $r' < r$. For the case $r > r'$, we can write,

$$\frac{1}{|\mathbf{r}-\mathbf{r}'|} = \sum_0^{\infty}\frac{r^l}{r'^{l+1}}P_l(\cos\theta)$$

As emphasised earlier, each term on the right-hand side is a solution of the Laplace's equation. The above expansions of $\frac{1}{|\mathbf{r}-\mathbf{r}'|}$ are found very useful in (i) multipole expansions (ii) boundary value problems in electrostatics.

Solved problem 4.4 Show using the generating function $g(x,t)$ that

(a) $P_n(1) = 1$

(b) $P_n(0) = 0$; for n odd and $P_{2n}(0) = \frac{(-)^n 2n!}{2^{2n}(n!)^2}$

(c) $P_n(-x) = (-)^n P_n(x)$

Solution:

(a) Recall that

$$g(x,t) = (1 - 2xt + t^2)^{-1/2} = \sum_{n=0}^{\infty} t^n P_n(x).$$

We set $x = 1$ in the above to obtain

$$g(1, t) = (1 - 2t + t^2)^{-1/2} = (1 - t)^{-1}$$

$$= \sum_{n=0}^{\infty} t^n = \sum_{n=0}^{\infty} t^n P_n(1).$$

Comparing coefficients of t^n, we obtain $P_n(1) = 1$.

(b) We now set $x = 0$ in the expression for the generating function to obtain

$$g(0, t) = (1 + t^2)^{-1/2}$$

$$= \sum_{n=0}^{\infty} t^{2n} \frac{(-)^n 2n!}{2^{2n}(n!)^2}$$

$$= \sum_{m=0}^{\infty} t^m P_m(0).$$

Comparing coefficients of t^m, we obtain $P_m(0) = 0$ for an odd m and $P_{2n}(0) = \frac{(-)^n 2n!}{2^{2n}(n!)^2}$.

(c) We now replace $x \to -x$ in the expression for the generating function to obtain

$$g(-x, t) = (1 + 2xt + t^2)^{-1/2} = g(x, -t).$$

Hence, $$\sum_{n=0}^{\infty} t^n P_n(-x) = \sum_{n=0}^{\infty} (-t)^n P_n(x).$$

A comparison of coefficients of t^n yields (c).

Solved problem 4.5 Use the recurrence relation

$$(n + 1)P_{n+1}(x) = (2n + 1)x P_n(x) - n P_{n-1}(x)$$

and the expressions for $P_n(x)$; $n = 0, 1, 2$ and 3 to evaluate (a) $P_4(x)$; (b) $P_6(0)$.

Solution:

(a) We set $n = 3$ in the above recurrence relation. We find that

$$4P_4(x) = 7xP_3(x) - 3P_2(x)$$
$$= 7x\left[\frac{1}{2}(5x^3 - 3x)\right] - 3\left[\frac{1}{2}(3x^2 - 1)\right].$$

This leads us to

$$P_4(x) = \frac{35x^4 - 30x^2 + 3}{8}.$$

(b) We set $n = 5$ and $x = 0$ in the recurrence relation to find

$$6P_6(0) = -5P_4(0) = -\frac{15}{8}.$$

Thus, $P_6(0) = -\frac{5}{16}$.

4.7 Associated Legendre Functions

As explained earlier in Chapter 1, when one separates the Laplace's equation or the Helmholtz wave equation in spherical polar coordinates, one ends up with a differential equation for the θ-dependent part of the solution that is given by

$$\frac{1}{\sin\theta}\frac{d}{d\theta}\left[\sin\theta\frac{dB(\theta)}{d\theta}\right] + \left[l(l+1) - \frac{m^2}{\sin^2\theta}\right]B(\theta) = 0.$$

If we change the variable to $x = \cos\theta$, we find that it can be rewritten as

$$(1 - x^2)y''(x) - 2xy'(x) + \left[l(l+1) - \frac{m^2}{1 - x^2}\right]y(x) = 0.$$

Here, we have expressed the dependent variable as $y(x) \equiv B(\theta)$. For $m = 0$, we recognise this as the Legendre differential equation. It can be shown that (see Exercise 1.1 of Chapter 1) in the case ($0 \le m \le l$), *one* of the solutions of the differential equation can be written as

$$y(x) = (1 - x^2)^{m/2}\frac{d^m}{dx^m}P_l(x).$$

We see, evidently, that the above definition is meaningful only when m is a nonnegative integer $\le l$.

It turns out that in most physical contexts, it is this solution that is relevant. The reason for this is based upon the fact that in physics we are interested, for physical reasons, in solutions that are single-valued and finite. As elaborated later (see Section 4.8), while discussing spherical harmonics, this requires that the m in the above equation is an integer. Further, as we have pointed out, the Legendre differential equation has solutions that are finite for *all* physical values of θ, viz., $0 \leq \theta \leq \pi$ only if l happens to be a nonnegative integer.

This solution, with the normalisation chosen below, is denoted by $P_l^m(x)$, the *associated Legendre functions*, and is defined as follows[18].

$$P_l^m(x) = (-)^m (1-x^2)^{m/2} \frac{d^m}{dx^m} P_l(x); \quad 0 \leq m \leq l \tag{4.33}$$

The Rodriguez formula allows one to re-express this as

$$P_l^m = (-)^m \frac{1}{2^l l!} (1-x^2)^{m/2} \frac{d^{l+m}}{dx^{l+m}} (x^2-1)^l. \tag{4.34}$$

The above formula for $P_l^m(x)$ can now be extended for negative values of m. In other words, with m nonnegative, we can write

$$P_l^{-m} = (-)^m \frac{1}{2^l l!} (1-x^2)^{-m/2} \frac{d^{l-m}}{dx^{l-m}} (x^2-1)^l. \tag{4.35}$$

Thus, henceforth, we will use the above definition (4.34) for all admissible m: $-l \leq m \leq l$

We shall now enumerate the properties of the associated Legendre functions.

(a) We note that for $m = 0$, the above definition reduces to that for the Legendre polynomials: $P_l^0(x) = P_l(x)$.

(b) We note that except in the case $m = 0$, the associated Legendre functions are *not* polynomials in x on account of the factor $(1-x^2)^{m/2}$. On the other hand, if we look upon $P_l^m(x)$ as a function of *two* (and *not* independent) variables $\cos\theta$ *and* $\sin\theta$, then $P_l^m(\theta)$ appears as a polynomial. This is so because $(1-x^2)^{m/2} = \sin^m\theta$ and $\frac{d^{l+m}}{dx^{l+m}}(x^2-1)^l$ are both so.

(c) $P_l^m(-x) = (-)^{l+m} P_l^m(x)$.

(d) In the physical context, we shall be using x as $\cos\theta$, where θ is the polar angle of the spherical polar coordinate system. Under parity operation, $\theta \to \pi-\theta$; $\phi \to \phi\pm\pi$ and thus, $x \to -x$. Then the observation (c) above can be reinterpreted as the statement that the parity of $P_l^m(x)$ is $(-)^{l+m}$.

(e) For $m > 0$, we have $P_l^m(\pm 1) = 0$. This is because $P_l^m(x)$ carries at least one factor of $(1-x^2)^{1/2}$. Further, in view of the relation between $P_l^m(x)$ and $P_l^{-m}(x)$ proven below, this property holds for $m < 0$ also.

[18] The factor of $(-)^m$ is a matter of convention. Definitions are often given without it.

(f) The associated Legendre functions satisfy

$$P_l^{-m}(x) = (-)^m \frac{(l-m)!}{(l+m)!} P_l^m(x). \tag{4.36}$$

Please note that the expressions for $P_l^{-m}(x)$ and $P_l^m(x)$ thus have a *different* normalisation.

Proof: We write d/dx as the differential operator D and write $X = x - 1$ and $Y = x + 1$. In this notation, we can write

$$2^l l! P_l^{-m} = (-)^m (1-x^2)^{-m/2} D^{(l-m)}(X^l Y^l)$$

and

$$2^l l! P_l^m = (-)^m (1-x^2)^{m/2} D^{l+m}(X^l Y^l).$$

Thus, proving the above result (4.36) requires that

$$D^{l-m}(X^l Y^l) = \frac{(l-m)!}{(l+m)!}(XY)^m D^{l+m}(X^l Y^l). \tag{4.37}$$

To prove this, we expand out both the sides by the use of the Leibniz theorem. We also use $D^q X^l = \frac{l!}{(l-q)!} X^{(l-q)}$ for $0 \le q \le l$ and a similar equation for Y. We thus find

$$\begin{aligned}
D^{l-m}(X^l Y^l) &= \sum_{q=0}^{l-m} \frac{(l-m)!}{q!(l-m-q)!} D^q X^l D^{l-m-q} Y^l \\
&\sum_{q=0}^{l-m} \frac{(l-m)!}{q!(l-m-q)!} \frac{l!}{(l-q)!} \times \frac{l!}{(m+q)!} X^{(l-q)} Y^{(m+q)} \\
&= (XY)^m \sum_{q=0}^{l-m} \frac{(l-m)!}{q!(l-m-q)!} \frac{l!}{(l-q)!} \times \frac{l!}{(m+q)!} X^{(l-m-q)} Y^q.
\end{aligned}$$

On the other hand, keeping in mind $D^{l+q} X^l = 0 = D^{l+q} Y^l$ for $q > 0$, we find

$$\begin{aligned}
D^{l+m}(X^l Y^l) &= \sum_{q=0}^{l-m} \frac{(l+m)!}{(m+q)!(l-q)!} D^{m+q} X^l D^{l-q} Y^l \\
&= \sum_{q=0}^{l-m} \frac{(l+m)!}{(m+q)!(l-q)!} \frac{l!}{(l-m-q)!} \frac{l!}{q!} X^{l-m-q} Y^q.
\end{aligned}$$

A comparison confirms (4.37), thus proving the relation (4.36) between P_l^{-m} and P_l^m.

(g) We shall write down the associated Legendre functions in a few simple cases. It is more useful to express these as functions of θ

$$P_0^0(x) = P_0(x) = 1$$

$$P_1^0(x) = P_1(x) = \cos\theta;$$

$$P_1^1(x) = -\sin\theta; \quad P_1^{-1}(x) = \frac{1}{2}\sin\theta$$

$$P_2^0(x) = P_2(x) = \frac{1}{2}\left(3\cos^2\theta - 1\right);$$

$$P_2^1(x) = -3\sin\theta\cos\theta; \quad P_2^{-1}(x) = \frac{1}{2}\sin\theta\cos\theta$$

$$P_2^2(x) = 3\sin^2\theta; \quad P_2^{-2}(x) = \frac{1}{8}\sin^2\theta$$

(h) We shall now prove the orthogonality property of the associated Legendre functions. It reads:

$$\int_{-1}^{1} dx\, P_l^m(x) P_{l'}^m(x) = \delta_{ll'} \frac{2}{2l+1} \frac{(l+m)!}{(l-m)!} \tag{4.38}$$

Note that the above relation involves the associated Legendre polynomials with the *same m*.
We shall first prove the orthogonality for $l \neq l'$. To this end, we note:

(i) For $m = 0$, the result follows immediately from the orthogonality property of the Legendre polynomials given earlier in Section 4.6.

(ii) We need to prove the result only for a positive m, because, then the result for the negative m follows immediately by the use of (4.36).

(iii) To prove this result for positive m, we need to show

$$\int_{-1}^{1} dx \left(1 - x^2\right)^m \frac{d^m}{dx^m} P_l(x) \frac{d^m}{dx^m} P_{l'}(x) = 0.$$

Without loss of generality, we shall assume that $l' < l$. We write the above integral as

$$\int_{-1}^{1} dx \left(1 - x^2\right)^m \frac{d^m}{dx^m} P_l(x) \frac{d^m}{dx^m} P_{l'}(x)$$

$$= (-)^m \int_{-1}^{1} dx\, P_l(x) \frac{d^m}{dx^m}\left[\left(1 - x^2\right)^m \frac{d^m}{dx^m} P_{l'}(x)\right]$$

$$= 0.$$

In going from the first step to the second, we have integrated by parts and then we have used the fact that $\frac{d^m}{dx^m}\left[(1-x^2)^m \frac{d^m}{dx^m}P_{l'}(x)\right]$, which is a polynomial of order $\leq l'$ (count powers of x) is orthogonal to $P_l(x)$ whose degree exceeds l'.

What then remains to be shown is the normalisation integral

$$\int_{-1}^{1} dx\left[P_l^m(x)\right]^2 = \frac{2}{2l+1}\frac{(l+m)!}{(l-m)!}$$

For, this purpose, we shall use the expression for $P_l^m(x)$ given by (4.34).

$$\begin{aligned}
&\int_{-1}^{1} dx\left[P_l^m(x)\right]^2 \\
&= \frac{1}{2^{2l}(l!)^2}\int_{-1}^{1} dx(1-x^2)^m \frac{d^{l+m}}{dx^{l+m}}(x^2-1)^l \frac{d^{l+m}}{dx^{l+m}}(x^2-1)^l \\
&= \frac{1}{2^{2l}(l!)^2}\int_{-1}^{1} dx(-)^{l+m}(x^2-1)^l \frac{d^{l+m}}{dx^{l+m}}\left[(1-x^2)^m \frac{d^{l+m}}{dx^{l+m}}(x^2-1)^l\right]
\end{aligned}$$

Here, we have first carried out integration by parts. We now note that the derivative $\frac{d^{l+m}}{dx^{l+m}}\left[(1-x^2)^m \frac{d^{l+m}}{dx^{l+m}}(x^2-1)^l\right]$ (which is an $(l+m)$-th derivative of a polynomial of degree $(l+m)$) is a number independent of x. It can be explicitly evaluated (see Exercise 4.23). Then the integral becomes

$$\frac{1}{2^{2l}(l!)^2}(-)^{l+m}(2l)!\frac{(l+m)!}{(l-m)!}\int_{-1}^{1} dx(x^2-1)^l.$$

The last integral can be performed using the beta function

$$\begin{aligned}
\int_{-1}^{1} dx(x^2-1)^l &= (-)^l\int_0^{\pi} d\theta\,\sin^{2l+1}\theta \\
&= (-)^l\frac{\Gamma(l+1)\Gamma(1/2)}{2\Gamma(l+3/2)} \\
&= (-)^l\frac{2^{2l+1}(l!)^2}{(2l)!},
\end{aligned}$$

leading to the result (4.38).

4.8 Spherical Harmonics

We discussed the separation of variables in the spherical polar coordinates for the Helmholtz wave equation $[(\nabla^2+k^2)\psi = 0]$ in Chapter 1. There we found that the functions that appear in the angular part of the solution of the form $B(\theta)C(\phi)$ satisfied the two ODE's, viz.,

$$\frac{1}{\sin\theta}\frac{d}{d\theta}\left[\sin\theta\frac{dB(\theta)}{d\theta}\right]+\left[l(l+1)-\frac{m^2}{\sin^2\theta}\right]B(\theta)=0 \qquad \text{(i)}$$

$$\frac{d^2C(\phi)}{d\phi^2}+m^2C(\phi)=0. \qquad \text{(ii)}$$

We note that the form of the equations is in fact independent of the presence of an r-dependent term in the original equation: we could consider, instead of the Helmholtz wave equation, the equation

$$(\nabla^2+k^2+U(r))\psi=0. \qquad (4.39)$$

An equation of this kind appears in the time-independent Schrödinger equation in presence of an arbitrary spherically symmetric potential. The form of the equations (i) and (ii) would, of course, apply to the Laplace's equation (set $k^2 = 0$ in the Helmholtz wave equation) and to the Poisson equation (with a (θ,ϕ)-independent source). Thus, the form for the angular part of the solution satisfying (i) and (ii) is of a very general applicability.

The solution to (ii) is of the form

$$C(\phi)=(\text{const})e^{\pm im\phi}.$$

When we separated the variables, we introduced a constant $(-m^2)$ (See equation (1.6) from Chapter 1) without placing any restriction on m. However, a restriction may arise in a given context from physical requirements. For example, we always require a physical observable to be a single-valued function[19]; so we must require $C(\phi=0)=C(\phi=2\pi)$. Thus, $C(\phi)$ is single-valued only if m is *real* and moreover an integer. This m also appears in (i), and we shall therefore consider (i) only with an integral m. We have further seen that (i) has as one of its solutions

$$B(\theta)=P_l^m(\cos\theta)$$

whenever l refers to a nonnegative integer greater than or equal to $|m|$. Henceforth, we shall consider (i) with only an integral value of l. The reasons for this have been elaborated in Section 4.7. Then we find that the angular part of the solution for (4.39) has the form:

$$B(\theta)C(\phi)=P_l^m(\cos\theta)e^{im\phi}$$

[19] In those cases where ψ of (4.39) refers to a wave function, it is strictly not obvious as to why single-valuedness be required for ψ. However, it proves to be generally sufficient.

We shall find the above combination in the context of all sorts of physical problems, be it the hydrogen atom (or the more complicated atomic systems and orbitals), or the potential scattering from a spherically symmetric potential, or the multipole expansion of an electromagnetic field, or the solutions to electrostatic problems[20]. This proves to be a complete set. We shall use it with a properly *normalised* form, the so-called 'spherical harmonics':

$$Y_{lm}(\theta,\phi) = \sqrt{\frac{2l+1}{4\pi}\frac{(l-m)!}{(l+m)!}}\, P_l^m(\cos\theta)e^{im\phi}.$$

As we shall later see, the factor involving the square root has been introduced for the sake of normalisation, while the factor $(-)^m$ present in $p_l^m(\cos\theta)$ is not necessitated by normalisation but by a matter of a commonly employed *convention* and is called the Condon–Shortley phase. From the way we have arrived at the spherical harmonics (see equations (i) and (ii)), it is clear that it satisfies

$$\nabla^2_{\text{ang}} Y_{lm} = -l(l+1)Y_{lm}, \tag{4.40}$$

where we have defined

$$\nabla^2_{\text{ang}} = \left[r^2\nabla^2 - \frac{\partial}{\partial r}\left(r^2\frac{\partial}{\partial r}\right)\right].$$

We shall now enumerate several simple properties of spherical harmonics.

(a) First, we note that $Y_{lm}(\theta,\phi)$ is a (generally) *complex* quantity. A physical observable will contain only real combinations of these. As seen below, its complex conjugate is simply related to another spherical harmonic $Y_{l,-m}(\theta,\phi)$.

(b) We also note that for $m = 0$, we have

$$Y_{l0}(\theta,\phi) = \sqrt{\frac{2l+1}{4\pi}}\, P_l^0(\cos\theta) = \sqrt{\frac{2l+1}{4\pi}}\, P_l(\cos\theta).$$

Thus, Y_{l0} is real and is proportional to the Legendre polynomial P_l.

(c) We note that

$$Y_{l,-m}(\theta,\phi) = \sqrt{\frac{2l+1}{4\pi}\frac{(l+m)!}{(l-m)!}}\, P_l^{-m}(\cos\theta)e^{-im\phi}$$

$$= \sqrt{\frac{2l+1}{4\pi}\frac{(l-m)!}{(l+m)!}}\,(-)^m P_l^m(\cos\theta)e^{-im\phi}$$

$$= (-)^m Y^*_{lm}(\theta,\phi).$$

Here we have used the property relating the P_l^{-m} to P_l^m of (4.36).

---

[20]Note that all of these relate to the partial differential equations mentioned earlier.

(d) We note that $Y_{lm}(\theta, \phi)$ has a definite parity $(-)^l$. For, under parity, $\theta \to (\pi - \theta)$ and $\phi \to (\pm\pi + \phi)$. Thus, $\cos\theta$ changes sign. Now, we recall that $P_l^m(-x) = (-)^{l+m} P_l^m(x)$. Hence,

$$Y_{lm}(\theta, \phi) \to Y_{lm}(\pi - \theta, \phi + \pi) = (-)^{l+m}(-)^m Y_{lm}(\theta, \phi) = (-)^l Y_{lm}(\theta, \phi).$$

(e) We note that for $\theta = 0$, $Y_{lm}(\theta, \phi)$ vanishes if $m \neq 0$ and is a constant independent of ϕ if $m = 0$. To see this, consider,

$$Y_{lm}(0, \phi) \propto P_l^m(1)$$

and

$$P_l^m(1) = \delta_{m0} P_l^0(1) = \delta_{m0} P_l(1) = \delta_{m0}.$$

These lead to

$$Y_{lm}(0, \phi) = \delta_{m0} Y_{l0}(0, \phi) = \delta_{m0}\sqrt{\frac{2l+1}{4\pi}}. \tag{4.40a}$$

(f) Finally, we consider the orthonormal property of the spherical harmonics,

$$\int_0^\pi \sin\theta d\theta \int_0^{2\pi} Y_{np}^*(\theta, \phi) Y_{lm}(\theta, \phi) d\phi \equiv \int d\Omega Y_{np}^*(\Omega) Y_{lm}(\Omega)$$
$$= \delta_{nl}\delta_{pm}. \tag{4.41}$$

The orthogonality in p and m follows from that of the phase factors $\exp(im\phi)$.

$$\int_0^{2\pi} \exp(im\phi)\exp(-ip\phi) = 2\pi\delta_{pm}$$

Then, we need to evaluate the left-hand side only for $p = m$. Now, if $p = m$, then the orthogonality in n and l follows from that for the associated Legendre functions for the *same* m as seen from (4.38). Finally, the normalisation in the definition of $Y_{lm}(\theta, \phi)$ has been so chosen as to ensure the ortho*normalily*, as seen by the use of (4.38).

(g) Just as a periodic function, say of ϕ, $0 \leq \phi < 2\pi$ can be expanded as a Fourier series in terms of $\sin m\phi$ and $\cos m\phi$ leading to successive 'harmonics', a function on unit sphere (rather than a unit circle) can be expanded in terms of the spherical harmonics:

$$f(\theta, \phi) = \sum_{l=0}^{\infty}\sum_{m=-l}^{l} a_{lm} Y_{lm}(\theta, \phi)$$

(h) We now enumerate a few of the simple spherical harmonics.

$l = 0$

$$Y_{00}(\theta, \phi) = \sqrt{\frac{1}{4\pi}} = \text{a constant}$$

$l = 1$

$$Y_{10}(\theta, \phi) = \sqrt{\frac{3}{4\pi}} P_1(\cos\theta) = \sqrt{\frac{3}{4\pi}} \cos\theta$$

$$Y_{11}(\theta, \phi) = -\sqrt{\frac{3}{8\pi}} \sin\theta e^{i\phi}$$

$$Y_{1,-1}(\theta, \phi) = -Y_{11}^*(\theta, \phi) = \sqrt{\frac{3}{8\pi}} \sin\theta e^{-i\phi}$$

$l = 2$

$$Y_{20}(\theta, \phi) = \sqrt{\frac{5}{4\pi}} P_2(\cos\theta) = \sqrt{\frac{5}{4\pi}} \frac{3\cos^2\theta - 1}{2}$$

$$Y_{21}(\theta, \phi) = -\sqrt{\frac{5}{24\pi}} 3\sin\theta\cos\theta e^{i\phi}$$

$$Y_{2,-1}(\theta, \phi) = -Y_{21}^*(\theta, \phi) = \sqrt{\frac{5}{24\pi}} 3\sin\theta\cos\theta e^{-i\phi}$$

$$Y_{22}(\theta, \phi) = \sqrt{\frac{5}{96\pi}} 3\sin^2\theta e^{2i\phi}$$

$$Y_{2,-2}(\theta, \phi) = \sqrt{\frac{5}{96\pi}} 3\sin^2\theta e^{-2i\phi}$$

It is possible to re-express these in a simple way in terms of the Cartesian coordinates. See the solved example below.

Solved problem 4.6 Express the spherical harmonics $r^l Y_{lm}(\theta, \phi) (0 \le l \le 2)$ given above in terms of the Cartesian coordinates.

Solution: An inspection of the results above and the relations

$$z = r\cos\theta; \ x \pm iy = r\sin\theta e^{\pm i\phi}$$

allow one to write

$$rY_{10}(\theta, \phi) = \sqrt{\frac{3}{4\pi}} z; \quad rY_{11}(\theta, \phi) = -\sqrt{\frac{3}{8\pi}}(x + iy);$$

$$rY_{1,-1}(\theta, \phi) = \sqrt{\frac{3}{8\pi}}(x - iy);$$

and,

$$r^2 Y_{20}(\theta, \phi) = \sqrt{\frac{5}{4\pi}} \frac{2z^2 - x^2 - y^2}{2}; \quad r^2 Y_{21}(\theta, \phi) = -\sqrt{\frac{5}{24\pi}} 3(x + iy)z$$

$$r^2 Y_{2,-1}(\theta, \phi) = -\sqrt{\frac{5}{24\pi}} 3(x - iy)z; \quad r^2 Y_{22}(\theta, \phi) = \sqrt{\frac{5}{96\pi}} 3(x + iy)^2;$$

$$r^2 Y_{2,-2}(\theta, \phi) = \sqrt{\frac{5}{96\pi}} 3(x - iy)^2.$$

Comments:

(a) We note that in each of the above cases, $r^l Y_{lm}$ is a polynomial in x, y, z of degree l.

(b) For $l = 1$, $r^1 Y_{1m}$ is directly proportional to (the linear combinations of) x, y and z which are components of a *vector*, i.e., an irreducible tensor operator of rank *one*. For $l = 2$, the five independent objects $r^2 Y_{2m}$ are precisely (linear combinations of) the five components of the irreducible tensor operator

$$\hat{T}_{ij} = \frac{1}{2}(T_{ij} + T_{ji}) - \frac{1}{3}\delta_{ij} T^k_k,$$

where $T_{ij} \equiv x_i x_j$, and is a traceless symmetric tensor of rank 2, i.e., same as l.

(c) The above observations are true in general for an arbitrary l. (see Exercise 4.25).

4.8.1 Addition theorem for spherical harmonics

Effect of rotation on $Y_{lm}(\theta, \phi)$

Consider a spherical harmonic $Y_{lm}(\Omega) \equiv Y_{lm}(\theta, \phi)$. It satisfies

$$\nabla^2_{\text{ang}} Y_{lm}(\theta, \phi) = -l(l + 1) Y_{lm}(\theta, \phi).$$

Now, we imagine rotating the coordinate frame about the origin. Then, θ and ϕ coordinates of a point will change but the operator $\nabla^2{}_{\text{ang}}$ (being a scalar operator) will remain unchanged. We shall denote the spherical harmonic in terms of new coordinates by $\tilde{Y}_{lm}(\theta', \phi') = Y_{lm}(\theta, \phi)$. It shall continue to satisfy

$$\nabla'^2_{\text{ang}} \tilde{Y}_{lm}(\theta', \phi') = -l(l + 1) \tilde{Y}_{lm}(\theta', \phi').$$

Now, the $(2l+1)$-dimensional vector subspace of eigenfunctions of $\nabla^2{}_{\text{ang}}$ with eigenvalue $-l(l+1)$ has the set $\{Y_{lm}(\Omega),\ m=-l,\ldots,l\}$ as the basis. Thus, we can express

$$\tilde{Y}_{lm}(\theta',\phi')=\sum_{n=-l}^{l}R_{mn}Y_{ln}(\theta,\phi).$$

Now (with $d\Omega\equiv\sin\theta d\theta d\phi$),

$$\int|Y_{lm}(\Omega)|^2d\Omega=\int|\tilde{Y}_{lm}(\theta',\phi')|^2d\Omega'=1=\int d\Omega|R_{mn}Y_{ln}(\theta,\phi)|^2.$$

(Note: The angular measure $d\Omega=d\Omega'$.) Using the orthogonality relation (4.41), this can be seen to imply that the $(2l+1)\times(2l+1)$ matrix R is unitary: $R^*_{mn}R_{mp}=\delta_{np}$.

Addition theorem for spherical harmonics

Let $\hat{\mathbf{r_1}}$ and Let $\hat{\mathbf{r_2}}$ be two unit vectors drawn from the origin characterised by the angular coordinates $\Omega_1\equiv(\theta_1,\phi_1)$ and $\Omega_2\equiv(\theta_2,\phi_2)$. Let Θ denote the angle between these two directions so that

$$\cos\Theta=\hat{\mathbf{r_1}}\cdot\hat{\mathbf{r_2}}=\cos\theta_1\cos\theta_2+\sin\theta_1\sin\theta_2\cos(\phi_2-\phi_1).$$

Now, consider

$$S\equiv\sum_{m=-l}^{m=l}Y^*_{lm}(\Omega_2)Y_{lm}(\Omega_1).$$

Under a rotation of coordinate axes,

$$S\to\sum_{m=-l}^{m=l}\sum_{n=-l}^{n=l}\sum_{p=-l}^{p=l}R^*_{mn}Y^*_{ln}(\Omega_2)R_{mp}Y_{lp}(\Omega_1)=\sum_{n=-l}^{l}Y^*_{ln}(\Omega_2)Y_{ln}(\Omega_1)=S,$$

where we have employed the unitarity of $R:R_{mp}R^*_{mn}=\delta_{np}$. Thus, S is invariant under rotations. We can therefore evaluate it with *any* choice of coordinate axes. Let us align the z-axis along $\hat{\mathbf{r_1}}$. Then, $\theta_1=0$ and by (4.40a), $Y_{lm}(0,\phi)=\delta_{m0}\sqrt{\frac{2l+1}{4\pi}}$. Thus,

$$S=\sqrt{\frac{2l+1}{4\pi}}Y^*_{l0}(\theta_2,\phi_2).$$

Now, we recall that

$$Y_{l0}(\theta_2,\phi_2)=\sqrt{\frac{2l+1}{4\pi}}P_l(\cos\theta_2)=Y^*_{l0}(\theta_2,\phi_2)$$

and with the present choice of coordinate axes, $\theta_2 = \Theta$. Thus, we find

$$P_l(\cos\Theta) = \frac{4\pi}{2l+1} S = \frac{4\pi}{2l+1} \sum_{m=-l}^{m=l} Y_{lm}^*(\Omega_2) Y_{lm}(\Omega_1). \tag{4.42}$$

This is referred to as the addition theorem for the spherical harmonics. It can be cast in a variety of forms. Noting that $P_l(\cos\Theta)$ is real, we can also write

$$P_l(\cos\Theta) = \frac{4\pi}{2l+1} \sum_{m=-l}^{m=l} Y_{lm}^*(\Omega_1) Y_{lm}(\Omega_2).$$

Recalling the relation between Y_{lm}^* and $Y_{l,-m}$, we can write

$$P_l(\cos\Theta) = \frac{4\pi}{2l+1} \sum_{m=-l}^{m=l} (-)^m Y_{l,-m}(\Omega_2) Y_{lm}(\Omega_1).$$

It can also be cast entirely in terms of real quantities as

$$P_l(\cos\Theta) = P_l(\cos\theta_1) P_l(\cos\theta_2) + 2\sum_{m=1}^{l} \frac{(l-m)!}{(l+m)!} P_l^m(\cos\theta_1) P_l^m(\cos\theta_2) \cos m(\phi_1 - \phi_2).$$

(see Exercise 4.27).

4.9 Laguerre Polynomials

In this section we shall summarise the results on the Laguerre polynomials and their relation to the hydrogen atom wave functions. Since much of the treatment is along the same lines as the Hermite and the Legendre polynomials, we shall be brief and leave much to exercises.

Summary of results regarding the Laguerre polynomials

1. **The Rodriguez formula**

$$L_n(x) = \frac{e^x}{n!} \frac{d^n}{dx^n} \left[e^{-x} x^n\right] \tag{4.43}$$

These are the special cases of the classical orthogonal polynomials over the interval $(0, \infty)$ found in Section 4.4, with $\rho = 0$. As seen from the above expression, the Laguerre polynomials do not have definite parity (except $n = 0$). This is unlike the Hermite and the Legendre polynomials which, unlike the present case, are moreover orthogonal polynomials over *symmetric* intervals.

2. **Orthogonality**
The Laguerre polynomials satisfy the orthogonality relation (see Exercise 4.30):

$$\int_0^\infty dx e^{-x} L_n(x) L_m(x) = \delta_{mn}.$$

3. **The generating function**
The Laguerre polynomials are generated by (see Exercise 4.31)

$$g(x,t) = \frac{e^{\frac{-xt}{1-t}}}{1-t} = \sum_0^\infty t^n L_n(x) \quad |t| < 1. \tag{4.44}$$

4. **The 3-term recurrence relation**
The Laguerre polynomials satisfy the 3-term recurrence relation of the form (4.18). It reads (see Exercise 4.32),

$$(n+1)L_{n+1}(x) = (2n+1-x)L_n(x) - nL_{n-1}(x).$$

(Note the presence of a nonvanishing $\beta_n x L_n$ term in this relation.)

5. **An additional recurrence relation** The Laguerre polynomials satisfy an additional recurrence relation characteristic to them.

$$xL_n'(x) = nL_n(x) - nL_{n-1}(x)$$

6. **The differential equation**
The Laguerre polynomials satisfy the second order linear ODE (see Exercise 4.32).

$$xL_n''(x) + (1-x)L_n'(x) + nL_n(x) = 0.$$

7. **A few examples**
The first few Laguerre polynomials are given by

$$L_0(x) = 1; \quad L_1(x) = 1 - x;$$

$$L_2(x) = \frac{1}{2}\left[x^2 - 4x + 2\right]; \quad L_3(x) = \frac{1}{6}\left[-x^3 + 9x^2 - 18x + 6\right].$$

(see Exercise 4.33).

4.9.1 The associated Laguerre polynomials

The associated Laguerre polynomials are also classical orthogonal polynomials over the internal $(0, \infty)$ with $w = e^{-x}x^k$, $k > -1$. They are given by the Rodriguez formula

$$L_n^k(x) = \frac{e^x x^{-k}}{n!}\frac{d^n}{dx^n}\left[e^{-x}x^{n+k}\right]. \tag{4.45}$$

They are also equivalently given by (see Exercise 4.35)

$$L_n^k(x) = (-)^k \frac{d^k}{dx^k}\left[L_{n+k}\right]. \tag{4.46}$$

The generating function for these can be written down from (4.44) by differentiation with respect to x, k times. It reads (see Exercise 4.34)

$$g_k(x,t) = \frac{\exp\left[\frac{-xt}{(1-t)}\right]}{(1-t)^{k+1}} = \sum_0^\infty t^n L_n^k(x); \quad |t| < 1. \tag{4.47}$$

The associated Laguerre polynomials satisfy the 3-term recurrence relation (see Exercise 4.36)

$$(n+1)L_{n+1}^k(x) = (2n+k+1-x)L_n^k(x) - (n+k)L_{n-1}^k(x),$$

an additional recurrence relation

$$xL_n^{k'}(x) = nL_n^k(x) - (n+k)L_{n-1}^k(x),$$

and the ODE

$$xL''^k_n + (k+1-x)L'^k_n(x) + nL_n^k(x) = 0.$$

They satisfy the orthogonality relation (see Exercise 4.37)

$$\int_0^\infty dx e^{-x} x^k L_n^k(x) L_m^k(x) = \frac{(n+k)!}{n!}\delta_{mn}. \tag{4.48}$$

The radial parts of the hydrogen atom wave functions are related to the associated Laguerre polynomials as follows: Let the wave function for the stationary state denoted by the quantum numbers (n, l, m) be expressed as

$$\psi(r,\theta,\phi) = R_{nl}(r)\Theta_{lm}(\theta,\phi) = R_{nl}(r)Y_{lm}(\theta,\phi).$$

Then, $R_{nl}(r)$ are given by

$$R_{nl}(r) = C_{lm} e^{-\alpha r/2}(\alpha r)^l L_{n-l-1}^{2l+1}(\alpha r),$$

where,

$$\alpha \equiv \frac{2me^2}{n\hbar^2}; \quad C_{lm} = \sqrt{\alpha^3 \frac{(n-l-1)!}{2n(n+l)!}}.$$

4.10 A Summary of Hypergeometric and Confluent Hypergeometric Functions

In this section we shall dwell briefly on the basic results regarding the hypergeometric and the confluent hypergeometric functions and their relation to the other special functions.

Consider the equation

$$z(z-1)y'' + [(a+b+1)z - c]y' + aby = 0. \tag{4.49}$$

This is called the hypergeometric equation. The equation has three regular singular points at $z = 0, 1, \infty$ (see Exercise 1.22 of Chapter 1). When $c \neq$ integer, the equation has two linearly independent solutions given by

$$y_1(z) = F(a, b|c|z),$$

$$y_2(z) = z^{1-c}F(a+1-c, b+1-c|2-c|z)), \tag{4.50}$$

where, $F(a, b|c|z)$ is been defined by the series

$$F(a, b|c|z) \equiv \sum_0^\infty \frac{(a+n-1)!(b+n-1)!(c-1)!}{(a-1)!(b-1)!(c+n-1)!(n)!}z^n \tag{4.51}$$

which converges in the domain $|z| < 1$ (see Exercise 1.22 of Chapter 1). This series is called the hypergeometric series or hypergeometric function.

4.10.1 Class of functions represented as hypergeometric series

1. *Geometric series*
   Set $a = c, b = 1$.

   $$F(a, 1|a|z) = 1 + z + z^2 + \cdots$$

2. *Logarithmic function*

   $$\frac{\ln(1+z)}{z} = F(1, 1|2|-z)$$

3. *Binomial series*

   $$(1+z)^a = F(-a, b|b|-z)$$

4. *Inverse sine function*

   $$\frac{\sin^{-1} z}{z} = F\left(\frac{1}{2}, \frac{1}{2}\middle|\frac{3}{2}\middle| z^2\right)$$

5. *Legendre polynomials*

$$P_n(x) = F\left(-n, n+1\middle|1\middle|\frac{1-x}{2}\right)$$

6. *Associated Legendre functions*

$$P_n^m(x) = \frac{(n+m)!}{(n-m)!}\frac{(1-x^2)^{m/2}}{2^m m!}F\left(m-n, m+n+1\middle|m+1\middle|\frac{1-x}{2}\right)$$

7. *Alternate forms of Legendre polynomials*

$$P_{2n}(x) = (-)^n\frac{(2n)!}{2^{2n}(n!)^2}F\left(-n, n+\frac{1}{2}\middle|\frac{1}{2}\middle|x^2\right)$$

$$P_{2n+1}(x) = (-)^n\frac{(2n+1)!}{2^{2n}(n!)^2}xF\left(-n, n+\frac{3}{2}\middle|\frac{3}{2}\middle|x^2\right)$$

Several other special functions can be represented in terms of hypergeometric functions.

4.10.2 Confluent hypergeometric function

Consider the ODE

$$zy'' + (c-z)y' - ay = 0. \tag{4.52}$$

This is called the confluent hypergeometric equation. When $c \neq$ integer, the two linearly independent solutions to this equation are:

$$\begin{aligned} y_1(z) &= F(a|c|z) \\ y_2(z) &= z^{1-c}F(a+1-c|2-c|z) \end{aligned} \tag{4.53}$$

where,

$$F(a|c|z) = \sum_0^\infty \frac{(a+n-1)!(c-1)!}{n!(c+n-1)!}z^n.$$

The series represents an entire function. Again, there is a large class of functions that can be represented by the confluent hypergeometric functions. For example:

1. *Hermite polynomials*

$$H_n(z) = (-)^{n/2}\frac{n!}{(n/2)!}F(-n/2|1/2|z^2) \quad n = 0, 2, 4, \ldots$$

$$H_n(z) = 2(-)^{(n-1)/2}\frac{n!}{[1/2(n-1)]!}zF\left(-n/2+\frac{1}{2}\middle|\frac{3}{2}\middle|z^2\right) \quad n = 1, 3, 5, \ldots$$

2. *Associated Laguerre functions* [and Laguerre polynomials ($\mu = 0$)]

$$L_n^{\mu}(z) = \frac{(n+\mu)!}{n!(\mu)!} F(-n|\mu+1|z)$$

3. *Error function*

$$\frac{\sqrt{\pi}}{2}\,\mathrm{erf}(z) \equiv \int_0^z \exp(-t^2)dt = zF\left(\frac{1}{2}\middle|\frac{3}{2}\middle| -z^2\right)$$

4. *Bessel functions*

$$J_\nu(z) = \frac{1}{\nu!}\left(\frac{z}{2}\right)^\nu \exp(-iz)F(\nu+1/2|2\nu+1|2iz)$$

4.11 Exercises

On § 4.4

4.1 Solve the equation determining $w(x)$ in each of the three cases and establish the general solutions given in Section 4.4.1

4.2 In Section 4.4.1, we solved for the possible weight functions for the cases of the three standard intervals. Verify that in each case the $C_n(x)$'s defined by (4.12) are in fact polynomials.

On § 4.5

4.3 Starting from the definition for $H_n(x) = (-)^n \exp(x^2)\frac{d^n}{dx^n}[\exp(-x^2)]$,

(a) Show that $H_n(-x) = (-)^n H_n(x)$.

(b) Show that the 3-term recurrence relation will not have the $\beta_n H_n$ term [Hint: Use your knowledge of parity of $H_n(x)$ in (a) above].

(c) Determine the coefficients of x^n and x^{n-1} in $H_n(x)$.

(d) Now starting from $H_{n+1}(x) - \alpha_n H_n(x) = \gamma_n H_{n-1}$ and using above values in (c), find γ_n.

(e) Construct $\sum_{n=0}^{\infty} \frac{t^n}{n!} H_n \equiv g(x,t)$ using the expression for H_n above and show that it equals $\exp(-t^2 + 2xt)$.

4.4 Verify the normalisation integral for the Hermite Polynomials:

$$\int_{-\infty}^{\infty} H_n^2(x)\exp(-x^2)\,dx = 2^n\sqrt{\pi}n!$$

4.5 Show that $H_n'(x) = 2nH_{n-1}(x)$. Do this using the generalised Rodriguez formula and also from the use of the generating function.

4.6 Show that the lowest power in $H_{2n}(x)$ is x^0 and its coefficient is $(-)^n \frac{(2n)!}{n!}$, while the lowest power in $H_{2n+1}(x)$ is x^1 and its coefficient is $\frac{2(-)^n(2n+1)!}{n!}$. [Hint: Use the generating function.]

4.7 Evaluate $\int_{-\infty}^{\infty} dx x H_n(x) H_m(x)$.

4.8 Show that all the zeros of $H_n(x)$ are real and distinct. [Hint: Use generalised Rodriguez formula and induction.]

On § 4.6

4.9 Derive the expression for $P_{2n}(0) = (-)^n \frac{(2n-1)!!}{(2n)!!}$.

4.10 Prove that $P_n'(1) = \frac{d}{dx} P_n(x)|_{x=1} = \frac{1}{2} n(n+1)$.

4.11 Evaluate $\int_{-1}^{1} P_n(x)dx$ and $\int_0^1 P_n(x)dx$

4.12 Determine the coefficients of x^n and x^{n-1} for the Legendre Polynomials $P_n(x)$.

4.13 Show that the coefficient of the highest power in $P_n(x)$ is $\frac{(2n)!}{n!}$ (Hint: Try the Rodriguez formula).

4.14 Show that the lowest power in $P_{2n}(x)$ is x^0 and its coefficient is $\frac{(-)^n[(2n)!]^2}{[n!]^2}$, while the lowest power in $P_{2n+1}(x)$ is x^1 and its coefficient is $\frac{(-)^n(2n+2)!(2n+1)!}{n!(n+1)!}$. [Hint: Use the Rodriguez formula.]

4.15 Show from an *inspection* of the generating function that $P_{2n+1}(0)$ vanishes and that $P_{2n}(0)$ alternates in sign.

4.16 Given (a) the knowledge of the parity of $P_n(x)$, (b) the normalisation condition $P_n(1) = 1$ and the orthogonality of $P_n(x)'s$ over $[-1, 1]$, construct $P_n(x)$ for $n = 0, 1, 2, 3$.

4.17 Show that the normalisation of the Legendre polynomials implied by the Rodriguez formula is equivalent to the one implied by $P_n(1) = 1$.

4.18 Find the value of the integral $\int_0^1 dx x^n P_l(x)$ for a fixed l and for any nonnegative integer n.

4.19 Starting from the Rodriguez formula, prove the normalisation integral

$$\int_{-1}^{1} [P_n(x)]^2 dx = \frac{2}{2n+1}.$$

4.20 Prove that (4.32) gives the generating function for the Legendre polynomials. Do it in the following steps:

(a) Expand $\left[1 - 2xt + t^2\right]^{-1/2}$ by the binomial theorem. Under what conditions on x and t can you do it?

(b) Expand $(2xt - t^2)^p$ by binomial theorem. Does it require any further restrictions?

(c) Pick out the coefficient of t^n in the expansion. Find the lowest and the highest powers of x in the expansion.

(d) Expand the expression for $P_n(x)$ given by the Rodriguez formula in powers of x. Show that the two results agree.

On § 4.7

4.21 Show that $P_n^n(\cos\theta) = (2n-1)!! \sin^n \theta$.

4.22 Evaluate $\int_0^1 dx\, P_n^n(x)$.

4.23 Evaluate $\frac{d^{l+m}}{dx^{l+m}}\left[(1-x^2)^m \frac{d^{l+m}}{dx^{l+m}}(x^2-1)^l\right]$ and show that it equals $\frac{(-)^m(2l)!(l+m)!}{(l-m)!}$. [Hint: Note that only the highest power of x in the square bracket contributes here. It is $(-x^2)^m \frac{d^{l+m}}{dx^{l+m}}(x^{2l})$.]

On § 4.8

4.24 Draw the locations on a unit sphere where the spherical harmonic $Y_{lm}(\theta,\phi)$ vanishes for $l = 2$ and $m = \pm 2, \pm 1, 0$.

4.25 Show that $r^l Y_{lm}(\theta,\phi)$ is a polynomial of degree l in x, y, z.

4.26 Quantum mechanical angular momentum operators $L_\pm$ are defined by

$$L_\pm = \pm e^{\pm i\phi}\left[\frac{\partial}{\partial\theta} \pm i\cot\theta \frac{\partial}{\partial\phi}\right].$$

Show that Y_{lm} satisfy

$$L_\pm Y_{lm} = \sqrt{(l \mp m)(l \pm m + 1)}\, Y_{l,(m\pm 1)}.$$

Show further that

$$Y_{lm} = \sqrt{\frac{(l+m)!}{(2l)!(l-m)!}}(L_-)^{l-m} Y_{ll} = \sqrt{\frac{(l-m)!}{(2l)!(l+m)!}}(L_+)^{l+m} Y_{l,-l}.$$

4.27 Show that the addition theorem for the spherical harmonics can alternately be cast as

$$P_l(\cos\Theta) = P_l(\cos\theta_1)P_l(\cos\theta_2) + 2\sum_{m=1}^{l}\frac{(l-m)!}{(l+m)!}P_l^m(\cos\theta_1)P_l^m(\cos\theta_2)\cos m(\phi_1-\phi_2).$$

4.28 A function may be expanded as a Laplace series

$$f(r,\theta,\phi) = \sum_{l,m} a_{lm} r^l Y_{lm}(\theta,\phi).$$

Show that angular average of $f(r,\theta,\phi)$ over a unit sphere is given by $f(0,0,0)$.

On § 4.9

4.29 Show that the leading term in $L_n(x)$ is $(-)^n x^n$.

4.30 Verify that the orthogonality relation for the Laguerre polynomials is a special case of (4.1). Also check normalisation using (4.43). [Hint: Use the form of $L_n(x) = (-x)^n +$ a polynomial of degree $\leq (n-1)$.]

4.31 Prove the generating function of the Laguerre polynomials as given by (4.44).

4.32 Use the generating function or the Rodriguez formula to derive the 3-term recurrence relation, the additional recurrence relation and the differential equation satisfied by the Laguerre polynomials.

4.33 Verify, using the generating function, the expressions for the first four Laguerre polynomials.

4.34 Starting from the generating function of (4.44), derive the generating function of the associated Laguerre polynomials of (4.47).

4.35 Derive the expression (4.45).

4.36 Staring from the generating function (4.47), obtain the 3-term recurrence relation, the additional recurrence relation and ODE for the associated Laguerre polynomials.

4.37 Verify the orthogonality relation (4.48). [Hint: Note that this is a special case of (4.1). To prove the normalisation, first obtain the leading term in $L_n^k(x)$.]

On § 4.10

4.38 Verify the results stated in Section 4.10 regarding the relation of the hypergeometric and the confluent hypergeometric functions with other known functions.

References

[1] Sansone, G., *Orthogonal Functions*, Interscience Publishers, New York, 1959.

[2] Szego, G., *Orthogonal Polynomials*, American Mathematical Society, New York, 1959.

5 Partial Differential Equations

5.1 Introduction

In Chapter 1, we had elaborated on the reasoning behind the fact that many of the fundamental laws of physics are formulated in terms of partial differential equations (PDE's). A partial differential equation[21] has more than one independent variable (typically x, y and z; or x, y, z and t in a physical context) and is an equation involving partial derivatives of a dependent variable, say ψ, with more than one independent variable. A second order partial differential equation involves some second order partial derivative of ψ as the highest order derivative of ψ.

Let us take a look at the set of partial differential equations xthat appear in physics.

1. The Laplace's equation (electrostatics, Newtonian gravity, hydrodynamics): $\nabla^2 \psi = 0$
2. The Poisson equation (electrostatics, Newtonian gravity): $\nabla^2\psi = -\rho$
3. The Helmholtz wave equation (various wave propagations and also the time-independent Schrödinger equation): $\nabla^2\psi + k^2\psi = 0$
4. The Schrödinger equation for a particle in presence of a potential $V(\mathbf{r})$:

$$\left(-\frac{\hbar^2}{2m}\nabla^2 + V(\mathbf{r})\right)\psi = i\hbar\frac{\partial}{\partial t}\psi$$

5. The time-independent Schrödinger equation:

$$\left(-\frac{\hbar^2}{2m}\nabla^2 + V(\mathbf{r}) - E\right)\psi = 0$$

6. The diffusion equation: $\frac{\partial}{\partial t}u - k\nabla^2 u = 0$
7. The wave equation: $\left[\nabla^2 - \frac{1}{c^2}\frac{\partial^2}{\partial t^2}\right]\psi = 0$
8. The Klein–Gordon equation: $\left[\nabla^2 - \frac{1}{c^2}\frac{\partial^2}{\partial t^2} - m^2\right]\psi = 0$
9. The Dirac equation: $\left[i\partial\!\!\!/ - m\right]\psi = 0$

[21] The exact definitions will be introduced in the next section.

We notice that all these equations are partial differential equations and most are of the second order. They contain 3 or 4 independent variables and involve partial derivatives of ψ with respect to more than one independent variable. Barring the Dirac equation, they do involve some second order partial derivatives and no derivatives of higher order and they all happen to contain the operator ∇^2.

In Chapter 1, we formulated a way of obtaining the general solution to such PDE's. It was based upon the method of separation of variables which we had introduced there. We had found that we could reduce such partial differential equations to a set of three ordinary differential equations. We had then developed methods to solve these. We found that the methods formulated there enable one to obtain a set of solutions each of which is of the separable form (1.3). We had further elaborated there that these solutions suffice to expand any solution to the physical problem at hand and thus constitute a complete set for this purpose. In short, the general solution to the above equations representing all possible physical situations is available through the method.

To understand the task that remains to be completed, we shall make a comparison with the case of a familiar ODE. We recall that the motion of a particle as described via the Newton's second law $\mathbf{F} = m\mathbf{a} = m\frac{d^2}{dt^2}\mathbf{r}$ is a set of ordinary differential equations. The integration of $\mathbf{F} = m\mathbf{a}$ gives us the *general* solution of the problem with six integration constants. The specific trajectory of a particle undergoing the motion is not known until appropriate conditions that determine the six constants are specified. In a problem such as this, where there is only one independent variable t, the answer is simple: one can (for example) specify the position and the velocity at the initial instant. It is not difficult to see that if a smaller set of initial conditions is specified, the trajectory will not be unique. If one specifies too much information on the other hand (for example the initial *and* the final positions and the initial velocity), one may not have a solution.

Similarly, in the case of the above PDE's we need to know what set of boundary conditions are appropriate for each problem. To appreciate the point that the nature of boundary conditions appropriate for a given PDE that can lead to a unique and physically meaningful solution can very much be dependent upon the nature of partial differential equations, we consider the following illustration.

Solved problem 5.1 Consider the differential equations

(a) $$\left[\frac{\partial^2}{\partial x^2} + \frac{\partial^2}{\partial y^2}\right]\psi = 0$$

(b) $$\left[\frac{\partial^2}{\partial x^2} - \frac{\partial^2}{\partial y^2}\right]\psi = 0$$

(c) $$\left[\frac{\partial^2}{\partial x^2} - \frac{\partial}{\partial y}\right]\psi - b = 0 \qquad b \neq 0$$

on and within a square of side 'a' with corners at $(0, 0)$, $(0, a)$, (a, a) and $(a, 0)$. Consider the boundary condition $\psi = 0$ on the sides of the square. Assume that the second order derivatives of ψ exist and are continuous on and within the square, if required. Show that the PDE (a) has a unique solution; the PDE (b) has an infinity of solutions; while the PDE (c) has no solution at all. Show further that the infinity of solutions in the case (b) do not exist if one side of the square is slightly increased to $(a + \epsilon)$ (to make it a rectangle).

Solution:

(a) We know that the solution to the Laplace's equation attains its maximum *as well as* minimum value on the boundary (see e.g., Chapter 7 of *Mathematical Physics: The Basics*). But the value of ψ on all of the boundary is zero. Thus, ψ must vanish throughout the interior too (clarify!). Thus the solution in this case is unique and is given by $\psi(x, y) = 0$ within the square.

(b) We note that $\psi = 0$ is always a solution of $\left[\frac{\partial^2}{\partial x^2} - \frac{\partial^2}{\partial y^2}\right]\psi = 0$ with the given boundary conditions. But one can also write down an infinite set of solutions such as

$$\psi = A_n \sin\frac{n\pi x}{a}\sin\frac{n\pi y}{a}; \quad n = 1, 2, 3, \ldots$$

(in the context of the wave equation with $y = ct$, we normally call these the standing modes on a string of length 'a'). Thus there are an infinite number of solutions in this case.

We further see that if one of the sides (say parallel to y-axis) is changed to b, then there is *no* solution to the PDE with the given boundary condition except $\psi = 0$. Note that the obvious modification one would expect:

$$\psi = A_n \sin\frac{n\pi x}{a}\sin\frac{n\pi y}{b}; \quad n = 1, 2, 3, \ldots$$

(while it does satisfy the boundary conditions) does not satisfy the PDE.

(c) We evaluate the PDE $\left[\frac{\partial^2}{\partial x^2} - \frac{\partial}{\partial y}\right]\psi - b = 0$ at the point[22] $(0,0)$. We find that it requires

$$\left(\frac{\partial^2}{\partial x^2} - \frac{\partial}{\partial y}\right)\psi(x, y)\bigg|_{(x,y)=(0,0)} = b \neq 0.$$

On the other hand, we find by inspection that the specified boundary conditions require

$$\frac{\partial^2}{\partial x^2}\psi(x, y)\bigg|_{(x,y)=(0,0)} = 0 = \frac{\partial}{\partial y}\psi(x, y)\bigg|_{(x,y)=(0,0)}.$$

[22] We note that the PDE is given to hold *on* and within the square. Hence the partial derivatives up to the second have been assumed to exist on the square.

This is clearly incompatible with the differential equation at (0, 0). This implies that there is *no* solution to this equation with this boundary condition.

The above example brings out how, given a common boundary condition, different *types* of differential equations can have totally different properties regarding the existence and uniqueness of solutions.

In the future sections, we shall pursue questions regarding the classification of boundary conditions and the PDE's and discuss results regarding the uniqueness and stability of solutions for combination types of PDE and boundary conditions.

5.2 Classification of Partial Differential Equations and Boundary Conditions

5.2.1 Some definitions

First, we shall introduce a number of notations and give several relevant definitions.

1. Let the number of independent variables of the differential equation be n; we shall denote these by $\{u_1, u_2, \ldots, u_n\}$. We shall denote any function of these, say $f(u_1, u_2, \ldots, u_n)$, by $f\{u_i\}$ for short.
2. The dependent variable of the differential equation will be denoted by $y(u_1, u_2, \ldots, u_n)$ or by $y\{u_i\}$ for short.
   We shall be dealing with *linear second order partial differential equations.* Below, we define these phrases.
3. **A partial differential equation** (PDE) has partial derivatives with respect to *at least two* independent variables present in it.
4. A **second order** PDE has some second order partial derivative (at least one) as the highest order derivative $\left[\text{e.g., } \frac{\partial^2 y}{\partial u_i \partial u_j}\right]$ present in it.
   We shall assume that only non-negative integral powers of partial derivatives of y occur in the PDE.
5. **Linearity:** Suppose one performs the scaling $y \to \alpha y$ in the PDE. If each term scales as α, then the equation is called 'linear homogeneous'. If, in addition to terms having linear scaling there are terms that do not change, the equation is said to be 'linear' (but not homogeneous).
   Boundary conditions for the PDE will be given on a hypersurface. We define the relevant quantities.
6. An $(n-1)$-dimensional **hypersurface** in the n-dimensional space whose points are $\{u_1, u_2, \ldots, u_n\}$ is defined by a condition of the form $f\{u_i\} = 0$. The **unit normal** to the surface is given by $\hat{n} = \frac{\nabla f}{|\nabla f|}$. Any unit vector $\hat{t}$ satisfying $\hat{t}.\hat{n} = 0$ is a unit vector tangent to the hypersurface.
7. For any function $y\{u_i\}$, the **normal derivative** $\frac{\partial y}{\partial n}$ at any point on the hypersurface is defined as $\frac{\partial y}{\partial n} = \hat{n}.\nabla y$. The $(n-1)$ linearly independent tangential derivatives

are obtained by evaluating $\hat{t}.\nabla y$ by choosing $(n-1)$ linearly independent tangent vectors.

In the subsequent discussion we shall assume, where required, that ψ has continuous second order derivatives in the region of interest.

Examples

1. Consider the Laplace's equation in 3 dimensions: $\nabla^2\psi = 0$. Here, the independent variables may be taken to be $\{x, y, z\}$ and $n = 3$. The dependent variable is $y \to \psi(x, y, z)$. The highest order of derivative present in it is the second. This is a *linear, homogeneous* PDE of *second* order.
2. In the context of the above equation, the following equation

$$f(x, y, z) \equiv x^2 + y^2 + z^2 - 1 = 0$$

defines a hypersurface (here a unit sphere). $\nabla f = 2x\hat{i} + 2y\hat{j} + 2z\hat{k}$ and $\hat{n} = \frac{\nabla f}{|\nabla f|} = \hat{r}$ given the unit normal to the surface. For any function $\psi(x, y, z)$, $\hat{r}.\nabla\psi = \frac{\partial\psi}{\partial r}$ gives the normal derivative of $\psi(r, \theta, \phi)$ at a point on the surface. $\hat{t}_1 = \hat{\theta}$ and $\hat{t}_2 = \hat{\phi}$ are two linearly independent unit tangents. $\hat{t}_1.\nabla\psi \equiv \frac{\partial\psi}{\partial\theta}$ and $\hat{t}_2.\nabla\psi \equiv \frac{\partial\psi}{\partial\phi}$ are the tangential derivatives.
3. The Poisson equation $\nabla^2\psi = \rho$ is a second order linear PDE that is *not* homogeneous.
4. The equation is

$$\sum_{i=1}^{3} \frac{\partial^2 y}{\partial {u_i}^2}\frac{\partial y}{\partial u_i} = u_i \frac{\partial y}{\partial u_i}.$$

This is a second order, but not linear, PDE.

5.2.2 Types of linear second order PDE's

We shall consider the most general PDE of second order (linear but not necessarily homogeneous). It reads

$$\sum_{i,j=1}^{n} a_{ij}\{u_p\}\frac{\partial^2 y}{\partial u_i \partial u_j} + F\left(u_i; y, \frac{\partial y}{\partial u_i}\right) = 0. \tag{5.1}$$

We shall assume that the function $y\{u_i\}$ encountered in any physical context has the required properties[23] such that each of the second ordered partial derivative $\frac{\partial^2 y}{\partial u_i \partial u_j}$ is symmetric in

[23] See e.g., *Mathematical Analysis* by Apostel, Second edition, Section 12.13. These properties are invariably met by the functions that occur in physics.

(i, j). Then, only the symmetric part of a_{ij} will contribute to the equation, and without loss of generality we can assume a_{ij} to be symmetric. Thus, at each point $(u_1, u_2, \ldots, u_n)$, $a_{ij}\{u\}$ is a real symmetric $n \times n$ matrix. Hence all the eigenvalues of the matrix are real. The PDE is classified according to the *signs* of these eigenvalues. The nomenclature very much resembles the nomenclature of the conic sections and their n-dimensional generalizations. A conic section is defined by a quadratic equation

$$a_{ij}x_i x_j + b_i x_i + c = 0,$$

and involves a 2×2 real symmetric matrix a_{ij} with eigenvalues λ_1 and λ_2. It represents an ellipse or a hyperbola or a parabola depending on whether both the eigenvalues (are nonzero and) have the same sign, distinct sign, or one of them vanishes. The quadratic equation then assumes the standard form:

(a)	An *Ellipse*	$\lambda_1 x'^2_1 + \lambda_2 x'^2_2 = 1$	$\lambda_i > 0$;
(b)	A *Hyperbola*	$\lambda_1 x'^2_1 + \lambda_2 x'^2_2 = 1$	$\lvert\lambda_i\rvert \neq 0,\ \lambda_1\lambda_2 < 0$;
(c)	A *Parabola*	$\lambda_1 x'^2_1 + b' x'_2 = 0$	

(Here, x'_1 and x'_2 are obtained by an appropriate rotation followed by a translation of origin, if necessary). We now come back to (5.1). At a given point, the $n \times n$ real symmetric matrix A of the coefficient functions a_{ij} can be diagonalised by an orthogonal transformation. If the coefficient functions $a_{ij}\{u\}$ are dependent on u_i's, then the eigenvalues can differ from point to point in the n-dimensional space. In such a case, we need to classify the PDE at a given point $\{u_i\}$.

We define the types of a PDE as follows: At point $\{u_i\}$, the equation is of the

(a) *elliptic type* if all the eigenvalues of A are *non-vanishing* and of the *same* sign;
(b) *hyperbolic type* if all the eigenvalues are *non-vanishing* but having *both* signs;
(c) *ultra-hyperbolic type* (for $n \geq 3$) if it is hyperbolic type and only *one* eigenvalue differs in sign from the rest;
(d) *parabolic type* if at least one of the eigenvalues *vanishes*.

Examples

1. Consider the examples considered earlier in Section 5.1:

(a) $\left[\dfrac{\partial^2}{\partial x^2} + \dfrac{\partial^2}{\partial y^2}\right]\psi = 0$

(b) $\left[\dfrac{\partial^2}{\partial x^2} - \dfrac{\partial^2}{\partial y^2}\right]\psi = 0$

(c) $\left[\dfrac{\partial^2}{\partial x^2} - \dfrac{\partial}{\partial y}\right]\psi - b = 0 \qquad b \neq 0$

In all these equations, we have two independent variables: x, y. These equations are already in a form such that the matrix $\{a_{ij}\}$ is diagonal. So, we can easily identify the type of the PDE: (a) is elliptic type, (b) is hyperbolic type, (c) is parabolic type.

2. In a similar manner, we find that
   (a) the Laplace's equation in 3 dimensions, $\nabla^2\phi = 0$, is of the elliptic type;
   (b) the wave equation
   $$\left[\frac{1}{c^2}\frac{\partial^2}{\partial t^2} - \nabla^2\right]\phi = 0$$
   is of the hyperbolic type, and, in fact, of an ultrahyperbolic type; and
   (c) the Schrödinger equation for a particle in a potential
   $$i\hbar\frac{\partial}{\partial t}\psi = \left[-\frac{\hbar^2}{2m}\nabla^2 + V(\mathbf{r})\right]\psi$$
   is a parabolic type equation.
3. Consider the PDE
   $$2xy\frac{\partial^2\psi}{\partial x\partial y} = \left(\frac{\partial\psi}{\partial x}\right)^2 + \left(\frac{\partial\psi}{\partial y}\right)^2.$$
   In this case we have two independent variables, x, y, and the 2×2 matrix of coefficients a_{ij} is: $\begin{bmatrix} 0 & xy \\ xy & 0 \end{bmatrix}$. The eigenvalues of this matrix are $\lambda_1 = xy$ and $\lambda_2 = -xy$. Thus we find that except at the origin, these are nonvanishing and of *opposite* signs. Thus the PDE is of the *hyperbolic* type.

5.2.3 Types of boundary conditions

A physical state of a system in (classical) physics is characterized by a set of unique values for its physical observables. For example, an electrostatic charge configuration experimentally leads to a unique electric field and a unique electrostatic potential (modulo a constant). A given source of electromagnetic waves produces a well-defined and measurable set of electromagnetic fields in space. One of the purposes of a formulation of a physical theory is to be able to predict these observed results from this formulation alone, given certain inputs about the physical system/physical situation. In a quantum mechanical system also, one may want to be able to calculate the wave function which is unique up to an *overall* phase. Now, in most cases, a physical theory is formulated in terms of the differential equations (Newton's equations of motion for the motion of a particle, Poisson equation in electrostatics, the wave equation for electromagnetic fields, etc.). As emphasised earlier in this chapter, these equations alone are not sufficient to yield a unique solution that corresponds to a *particular* physical situation. Additional information about a particular physical situation has to be supplied via 'boundary conditions' or initial conditions. If the boundary conditions provided are *appropriate* for the PDE in question, then the PDE together with these boundary conditions should be able to give a *unique* solution that can be compared with the experimentally observed (*and therefore, in principle, unique*) quantities. Further, we generally observe that[24] small changes in a physical system generally lead to small changes in observables. Hence we expect that the solution of the PDE so obtained must also be what is called '*stable*' with respect to small perturbations in boundary conditions. In this chapter we are going to formulate what kind of boundary conditions are suitable for what type of an PDE in the sense of yielding a **unique**, **stable** solution.

In this section we shall therefore first summarise the types of boundary conditions that *can*, in principle, be specified for a second order PDE under consideration. There are the following possibilities:

There are two kinds of boundaries on which a boundary condition is specified. It is either an open or a closed hypersurface in an n-dimensional space. The boundary conditions are further classified according to the information provided on the boundary.

We note that if we specify the value of y on a hypersurface, this information enables us to evaluate all the (linearly independent) derivatives of y *tangential* to the hypersurface (closed or open). Conversely, if all the linearly independent tangential derivatives of y are given on a hypersurface, it enables us to evaluate y on it (up to a constant) (see Exercise 5.16). The latter therefore does not constitute independent data. What *can* thus be specified independently is the derivative of y *normal* to the hypersurface.

(a) One can specify the *value* of the function $y\{u_i\}$ on a hypersurface. This is referred to as the **Dirichlet** boundary condition.
(b) One can specify the *normal* derivative of y, viz., $\frac{\partial y}{\partial n}$ on a hypersurface. This is referred to as the **Neumann** boundary condition.
(c) One can specify the *normal* derivative of y on a part of the hypersurface and y on the remaining part of the hypersurface. This is referred to as the **mixed** boundary condition[25].
(d) One can specify *both* the y and the $\frac{\partial y}{\partial n}$ on the hypersurface. This constitutes the **Cauchy** boundary condition.

Examples

(a) Consider the case of 3 independent variables $\{x, y, z\}$. Then, $x^2 + y^2 + z^2 = 1$ defines a closed hypersurface S (a unit sphere), whereas $x^2 + y^2 + z^2 = 1, z > 0$ defines an open hypersurface S_1 (a hemisphere). The complementary hypersurface $x^2 + y^2 + z^2 = 1, z \leq 0$ defines an open hypersurface S_2 (a hemisphere together with its boundary).

The examples (b) and (c) below use the notations introduced in this one.

[24] In this chapter, we are dealing with linear PDE's, and shall ignore non-linear phenomena.

[25] There can be another type of a mixed boundary condition in which one specifies $\frac{\partial y}{\partial n} + \alpha y$ on an hypersurface. We shall not deal with it in this book.

(b) Giving $y|_S = 0$ constitutes a *Dirichlet* boundary condition on a *closed* hypersurface. Given this boundary condition, it implies that all the tangential derivatives (here the two linearly independent ones $\frac{\partial y}{\partial \theta}, \frac{\partial y}{\partial \phi}$) of y vanish on S. Conversely, given that all tangential derivatives of y vanish on S, it follows that $y =$ constant on S. Thus, y is determined up to a constant.

(c) Suppose we specify $\frac{\partial y}{\partial r}\Big|_{S_1} = a(\theta, \phi)$. This constitutes a *Neumann* boundary condition on an *open* hypersurface.

(d) Giving $\frac{\partial y}{\partial r}\Big|_{S_1} = b(\theta, \phi)$ together with $y|_{S_1} = f(\theta, \phi)$ constitutes a *Cauchy* boundary conditions on an *open* hypersurface.

(e) Giving $\frac{\partial y}{\partial r}\Big|_{S_1} = c(\theta, \phi)$ together with $y|_{S_2} = g(\theta, \phi)$ corresponds to *mixed* boundary conditions on a *closed* hypersurface.

5.3 General Method: Two Dimensions

In this section we shall discuss the (as defined below) in two dimensions. Suppose we are given a second order linear PDE in two dimensions (x, y) with the dependent variable $\psi(x, y)$ which satisfies the PDE of the form

$$A(x, y)\psi_{xx} + 2B(x, y)\psi_{xy} + C(x, y)\psi_{yy} = D(\psi_x, \psi_y, \psi; x, y). \tag{5.2}$$

Suppose we are given the Cauchy boundary conditions on a smooth[26] curve C. Thus, on C we have been given

$$\psi(x, y)|_C = a(x, y) \quad \text{and} \quad \frac{\partial \psi(x, y)}{\partial n}\Big|_C = b(x, y).$$

We would like to discuss if it is possible to determine ψ in the vicinity of a given point (x_0, y_0) on C. With this in mind, we shall summarise what information is contained in the Cauchy boundary conditions given above. Let $\hat{\mathbf{t}}$ be the unit tangent at a point on C. Given $\psi(x, y)$ on C, we know the tangential derivative of ψ, viz., $\hat{\mathbf{t}} \cdot \nabla\psi$.

The tangential and the normal derivatives constitute two linearly independent derivatives in two dimensions, and thus from these we can deduce both ψ_x and ψ_y on C at each point. Further, assuming that the boundary conditions $a(x, y)$ and $b(x, y)$ are infinitely differentiable along the curve, we know all-order *tangential* derivatives of ψ_x and ψ_y and these again constitute certain linear combinations of successive higher order partial derivatives starting with the

[26] The discussion below generalizes easily to a piecewise smooth curve. The definition of a smooth curve is given for reference on page 162.

second. Now, the problem consists in finding if and under what conditions one can obtain the solution in the vicinity of the curve C.

The method presented below is applicable if there is a solution that is *analytic* in the two variables (x, y)[27]. By analyticity in (x, y), we refer to the existence of the Taylor expansion the function convergent in some neighbourhood in two dimensions. To be precise:

Definition: A function of N real variables $\psi(x_1, x_2, \ldots, x_N)$ is analytic at a point $(x_{10}, x_{20}, x_{30}, \ldots, x_{N0})$ if it can be expanded as a Taylor expansion,

$$\psi\{x_i\} = \psi\{x_{i0}\} + \sum_{j=1}^{N} \left.\frac{\partial \psi}{\partial x_j}\right|_{x_i=x_{i0}} (x_j - x_{j0}) + \cdots$$

$$+ \sum_{n_1, n_2, \ldots, n_N}^{n=\sum n_i} \frac{n!}{n_1! n_2! \cdots n_N!} \left.\frac{\partial^n \psi}{\partial^{n_1} x_1 \partial^{n_2} x_2 \ldots \partial^{n_N} x_N}\right|_{x_i=x_{i0}}$$

$$\times (x_1 - x_{10})^{n_1} \ldots (x_N - x_{N0})^{n_N} + \cdots$$

which converges in some neighbourhood of $\{x_{i0}\}$, say $|x_i - x_{i0}| < \epsilon_i > 0$. Thus, in the case of two independent variables (x, y), the above expansion reads,

$$\psi(x, y) = \psi(x_0, y_0) + \psi_x(x - x_0) + \psi_y(y - y_0) + \frac{1}{2}\psi_{xx}(x - x_0)^2$$

$$+ \psi_{xy}(x - x_0)(y - y_0) + \frac{1}{2}\psi_{yy}(y - y_0)^2 + \cdots$$

$$= \sum_{n_1, n_2=0}^{\infty} \left.\frac{\partial^{n_1+n_2}\psi}{\partial x^{n_1} \partial y^{n_2}}\right|_{(x_0, y_0)} \frac{(n_1 + n_2)!}{n_1! n_2!} (x - x_0)^{n_1} (y - y_0)^{n_2}, \tag{5.3}$$

where ψ_x etc. refer to the respective derivatives at (x_0, y_0). Now, we can determine an analytic solution in the vicinity of a point (x_0, y_0) on C if the coefficients of expansion, which are various order derivatives of ψ, can be determined at (x_0, y_0). This is the problem we shall address to, systematically.

---

[27] We shall use this term with a meaning different from what is used in the context of a function of a complex variable. This definition is valid for any number (n) of *real* variables. When applied to the case of *two* real variables, it is *less* restrictive than the definition of an analytic function of z. Thus, for example, $f(x, y) = \sin(xy)$ is analytic in the former sense but not in the latter sense. See Chapter 7 of *Mathematical Physics: The Basics* for more on this.

Let the smooth curve be parametrised by s, the distance measured on it from some fixed point (with $s \in (a, b)$), as[28]

$$x = X(s), \; y = Y(s), \quad a < s < b.$$

Then, the unit tangent and the unit normal to the curve at an arbitrary point $s = s_0$ are

$$\hat{\mathbf{t}} = \frac{d\mathbf{r}/ds}{|d\mathbf{r}/ds|} = X'(s)\hat{\mathbf{i}} + Y'(s)\hat{\mathbf{j}}, \tag{5.4}$$

$$\hat{\mathbf{n}} = Y'(s)\hat{\mathbf{i}} - X'(s)\hat{\mathbf{j}}. \tag{5.5}$$

The knowledge of $\psi(s)$ in the vicinity of (x_0, y_0), parameterised by $s = s_0$, allows us to evaluate

$$\psi'(s_0) = \hat{\mathbf{t}} \cdot \nabla\psi = X'(s_0)\psi_x + Y'(s_0)\psi_y$$

and the knowledge of $\frac{\partial \psi}{\partial n}$ at (x_0, y_0) allows us to evaluate another linearly independent combination of the first order derivatives

$$\frac{\partial \psi}{\partial n} = Y'(s_0)\psi_x - X'(s_0)\psi_y.$$

Thus, we find both the first order derivatives ψ_x and ψ_y at $s = s_0$. This applies to every point on C with $s \in (a, b)$.

Now, we turn to the second derivatives ψ_{xx}, ψ_{xy} and ψ_{yy}. We note that we have three sources of information about these.

(a) The PDE in which the values of ψ and the first order derivatives ψ_x, ψ_y at the point $s = s_0$ are substituted

(b) Knowledge of $\frac{\partial}{\partial s}\psi_x(s)$ (from the already determined $\psi_x(x(s), y(s))$ for $a < s < b$)

(c) Knowledge of $\frac{\partial}{\partial s}\psi_y(s)$ (from the already determined $\psi_x(x(s), y(s))$ for $a < s < b$)

Given these, the aim now is to find under what conditions it is possible to extract all the second order derivatives of ψ at $s = s_0$. We write down the equations obtained from the above three sources:

$$\begin{aligned} A(x, y)\psi_{xx} + 2B(x, y)\psi_{xy} + C(x, y)\psi_{yy} &= D(\psi_x, \psi_y, \psi, x, y) \\ &= \text{a known quantity} \end{aligned}$$

$$X'(s_0)\psi_{xx} + Y'(s_0)\psi_{xy} = \frac{\partial}{\partial s}\psi_x\bigg|_{s=s_0} = \text{a known quantity}$$

$$X'(s_0)\psi_{xy} + Y'(s_0)\psi_{yy} = \frac{\partial}{\partial s}\psi_y\bigg|_{s=s_0} = \text{a known quantity} \tag{5.6}$$

[28] A smooth curve has $X'(s)$ and $Y'(s)$ existing and continuous.

These can together be written as a matrix equation.

$$\begin{bmatrix} A & 2B & C \\ X'(s_0) & Y'(s_0) & 0 \\ 0 & X'(s_0) & Y'(s_0) \end{bmatrix} \begin{bmatrix} \psi_{xx} \\ \psi_{xy} \\ \psi_{yy} \end{bmatrix} = \text{a known column-vector.} \tag{5.7}$$

This equation can be solved for the three derivatives at $s = s_0$ if the determinant

$$\Delta \equiv \begin{vmatrix} A & 2B & C \\ X'(s_0) & Y'(s_0) & 0 \\ 0 & X'(s_0) & Y'(s_0) \end{vmatrix} \neq 0. \tag{5.8}$$

To proceed further, we shall assume that $\Delta \neq 0$ is satisfied at each point of some neighbourhood of the point $s = s_0$ on the curve (e.g., $|s - s_0| < \delta$) and all the second order derivatives have been determined at every point in this neighbourhood.

We shall then proceed to find how the third order partial derivatives can be evaluated. There are four of these: $\psi_{xxx}, \psi_{xxy}, \psi_{xyy}$ and ψ_{yyy}. Again, there are the following sources which can give us the information about these:

(a) The two first order derivatives of the PDE with respect to x and y in which the values of ψ, the first order derivatives ψ_x, ψ_y, and the second order derivatives ψ_{xx}, ψ_{xy} and ψ_{yy} at the point $s = s_0$ are substituted.

(b) The knowledge of the derivatives of ψ_{xx}, ψ_{xy} and ψ_{yy} along the curve.

Let us first consider the PDE of (5.2). We can differentiate[29] it with respect to, say x. We then obtain an equation of the form

$$A(x, y)\psi_{xxx} + 2B(x, y)\psi_{xxy} + C(x, y)\psi_{xyy} = \text{a known quantity.} \tag{5.9}$$

The above equation involves *three* out of four third order derivatives. We seek additional relations involving *these* three. We note that this is possible if we differentiate the *known* quantities ψ_{xx} and ψ_{xy} along the curve at $s = s_0$. This is much the same as done for the first order partial derivatives. We then have,

$$X'(s_0)\psi_{xxx} + Y'(s_0)\psi_{xxy} = \frac{\partial}{\partial s}\psi_{xx}\bigg|_{s=s_0} = \text{a known quantity,} \tag{5.10}$$

$$X'(s_0)\psi_{xxy} + Y'(s_0)\psi_{xyy} = \frac{\partial}{\partial s}\psi_{xy}\bigg|_{s=s_0} = \text{a known quantity.} \tag{5.11}$$

[29]The differential equation is valid in a two-dimensional *neighbourhood* of (x_0, y_0), and hence unlike the boundary conditions, which can be differentiated only *along* the curve C, it can be differentiated with respect to (x, y) independently.

We find that these three equations (5.9)–(5.11) can be solved for the three partial derivatives, provided,

$$\Delta \equiv \begin{vmatrix} A & 2B & C \\ X'(s_0) & Y'(s_0) & 0 \\ 0 & X'(s_0) & Y'(s_0) \end{vmatrix} \neq 0. \tag{5.12}$$

This condition is identical to the condition we had formulated earlier in the context of the second order derivatives. Assuming that this is fulfilled at $s = s_0$, we can solve for these equations. Further, any one of the additional equations involving the remaining derivative, ψ_{yyy}, viz.,

$$A(x, y)\psi_{xxy} + 2B(x, y)\psi_{xyy} + C(x, y)\psi_{yyy} = \text{a known quantity}, \tag{5.13}$$

$$X'(s_0)\psi_{xyy} + Y'(s_0)\psi_{yyy} = \frac{\partial}{\partial s}\psi_{yy}\bigg|_{s=s_0} = \text{a known quantity}, \tag{5.14}$$

will enable one to determine ψ_{yyy}. (How would one conclude that the answers one gets from (5.13) and (5.14) are identical? (See Exercise 5.14).

It can be verified that determination of the higher order derivatives also becomes possible if the condition (5.12) is fulfilled (see Exercise 5.10).

Thus, all the derivatives of $\psi(x, y)$ can be obtained at any point where the determinant $\Delta \neq 0$.

Solved problem 5.2 Consider the wave equation ($c = 1$)

$$\frac{\partial^2\psi}{\partial t^2} - \frac{\partial^2\psi}{\partial x^2} = 0$$

together with the following Cauchy boundary conditions:

$$\psi(x, 0) = 0, \quad \frac{\partial\psi}{\partial t}(x, 0) = 0, \quad -1 < x < 1.$$

Verify that one can uniquely evaluate all the partial derivatives of $\psi(x, t)$ at (say) the origin.

Solution: From the value of ψ in the interval $(-1, 1)$, we obtain

$$\frac{\partial^n\psi}{\partial x^n}(x, 0) = 0, \quad -1 < x < 1, \quad n = 1, 2, \ldots$$

From the value of $\frac{\partial\psi}{\partial t}$ in the neighbourhood of $(-1, 1)$, we obtain

$$\frac{\partial^{n+1}\psi}{\partial^n x\partial t} = 0, \quad -1 < x < 1 \quad n = 1, 2, \ldots \tag{5.15}$$

We now employ the wave equation: $\psi_{xx} = \psi_{tt}$ to obtain

$$\begin{aligned}\psi_{tt}(x, 0) &= \psi_{xx}(x, 0) = 0, \quad -1 < x < 1,\\ \psi_{ttx}(x, 0) &= \psi_{xxx}(x, 0) = 0, \quad -1 < x < 1,\\ \psi_{ttxx}(x, 0) &= \psi_{xxxx}(x, 0) = 0, \quad -1 < x < 1.\end{aligned} \tag{5.16}$$

We differentiate the wave equation $\psi_{tt} = \psi_{xx}$ with respect to t and obtain

$$\psi_{ttt} = \psi_{txx},$$

implying, from (5.15),

$$\psi_{ttt}(x, 0) = \psi_{txx}(x, 0) = 0, \quad -1 < x < 1.$$

One can continue along these lines and prove that *all* the partial derivatives of $\psi(x, t)$ on the curve $t = 0, -1 < x < 1$ can be determined and are zero.

5.3.1 Characteristics

We have seen that the Cauchy problem can give us an analytic solution in the vicinity of each point on the curve C provided $\Delta \neq 0$ at any point on it. In order to understand its implications for the curve C, we shall first consider what the condition $\Delta = 0$ implies.

We evaluate Δ and find that

$$\Delta \equiv Y'(s_0)^2 A - 2X'(s_0)Y'(s_0)B + X'(s_0)^2 C = 0. \tag{5.17}$$

We note that the above equation at $s = s_0$ relates to the *slope* of the curve C at (x_0, y_0), viz., $\frac{dy}{dx}\Big|_{(x_0,y_0)} = \frac{Y'(s_0)}{X'(s_0)}$. We can rewrite it as an equation determining the slope.

$$\left(\frac{dy}{dx}\right)^2 A - 2\frac{dy}{dx}B + C = 0 \tag{5.18}$$

This quadratic equation can be solved for the slope $\frac{dy}{dx}$ to yield

$$\frac{dy}{dx} = \frac{B \pm \sqrt{B^2 - AC}}{A}. \tag{5.19}$$

For each of the signs, the above equation determines a differential equation to be solved. The solution to each of the two equations contains an arbitrary constant and thus would, a priori, lead to two families of curves called *characteristics or characteristic curves.* A priori, there would be two characteristics that pass through (x_0, y_0), should they be real and distinct.

We make several observations.

(a) For a hyperbolic type of an equation, $B^2 - AC > 0$ and thus the square root appearing in (5.19) is in fact real. Thus, the characteristics are actually in the x–y plane and thus physically meaningful.

(b) For an elliptic type of an equation, $B^2 - AC < 0$ and thus the square root appearing in (5.19) is purely imaginary. The solutions to the equations then do not lie in the real space (i.e., x–y plane).

(c) For a parabolic equation, $B^2 - AC = 0$ and thus there is only one equation to the characteristics.

(d) From the above it is clear that the usefulness of the methods based on characteristics is relatively limited to the hyperbolic type PDE. For the elliptic type we shall need alternate methods, though the characteristics are useful in a formal sense.

To summarise, we note that all the coefficients of the Taylor expansion can be determined at a point on C if the curve C is not parallel to a characteristic through this point. **Assuming that the solution is analytic, the solution can then be determined in some neighbourhood of each point on C if C is nowhere parallel to a characteristic.**

Solved problem 5.3 Write down the equations of the characteristics through (x_0, y_0) and the *family* of characteristics for the PDEs below.

(a) $\left[\frac{\partial^2}{\partial x^2} + \frac{\partial^2}{\partial y^2}\right]\psi = 0$

(b) $\left[\frac{\partial^2}{\partial x^2} - \frac{\partial^2}{\partial y^2}\right]\psi = 0$

(c) $\left[\frac{\partial^2}{\partial x^2} - \frac{\partial}{\partial y}\right]\psi - b = 0 \qquad b \neq 0$

Solution:

(a) Here, $A = C = 1$, $B = 0$ and $\sqrt{B^2 - AC} = -1 < 0$. Thus, this is an elliptic equation. The equations for the characteristics are

$$\frac{dy}{dx} = \pm i.$$

The solutions to the family of characteristics are:

$$y + ix = C_1, \quad y - ix = C_2,$$

where C_1 and C_2 are arbitrary constants. The characteristics through (x_0, y_0) are

$$y - y_0 = \pm i(x - x_0).$$

(b) In this case, $A = -C = 1$, $B = 0$, $\sqrt{B^2 - AC} = 1 > 0$. Thus, this is a hyperbolic equation. The equations for the characteristics are

$$\frac{dy}{dx} = \pm 1.$$

The solutions to the family of characteristics are:

$$y + x = C_1, \quad y - x = C_2,$$

where C_1 and C_2 are arbitrary constants. The characteristics through (x_0, y_0) are:

$$y - y_0 = \pm(x - x_0).$$

(c) Here, $A = 1$, $B = C = 0$. Thus, $\sqrt{B^2 - AC} = 0$ and this is parabolic equation. There is only one equation for the characteristics and it is

$$\frac{dy}{dx} = 0$$

with the obvious solution, $y =$ constant. The characteristics through (x_0, y_0) is $y = y_0$.

For obvious reasons, we shall mainly concentrate on characteristics for the hyperbolic type equations. In this case the equations for characteristics are simple if A, B, C are all constants. Then the characteristics are families of straight lines.

$$y = m_1 x + c_1,$$

$$y = m_2 x + c_2,$$

$$m_1 = \frac{B + \sqrt{B^2 - AC}}{A}, \quad m_2 = \frac{B - \sqrt{B^2 - AC}}{A}, \tag{5.20}$$

with c_1, c_2 as arbitrary constants.

5.3.2 The partial differential equations in terms of the variables based on characteristics

We shall restrict ourselves with the PDE's in the two dimensions that have constant coefficients A, B and C. This includes the wave equation, the Poisson equation, the Laplace's equation, the Helmholtz wave equation in two dimensions.

We note the form of the family of characteristics in (5.20). We define the variables

$$\xi(x, y) = y - m_1 x, \quad \eta(x, y) = y - m_2 x. \tag{5.21}$$

Then the families of characteristics read as

$$\xi(x, y) = c_1, \quad \eta(x, y) = c_2. \tag{5.22}$$

We can cast the differential the PDE's in terms of the variables $\xi(x, y)$ and $\eta(x, y)$ or equivalently in terms of

$$X(x, y) = \frac{\xi(x, y) - \eta(x, y)}{2i}; \quad Y(x, y) = \frac{\xi(x, y) + \eta(x, y)}{2}. \tag{5.23}$$

It is easy to show that the PDE (5.2) can be cast in the following standard forms:

(i) The hyperbolic equation

$$\frac{\partial^2 \psi}{\partial \xi \partial \eta} = D'[\psi, \psi_\xi, \psi_\eta, \xi, \eta],$$

or equivalently,

$$\frac{\partial^2 \psi}{\partial X^2} - \frac{\partial^2 \psi}{\partial Y^2} = D'[\psi, \psi_X, \psi_Y, X, Y]$$

(ii) The elliptic equation

$$\frac{\partial^2 \psi}{\partial X^2} + \frac{\partial^2 \psi}{\partial Y^2} = D'[\psi, \psi_X, \psi_Y, X, Y]$$

(iii) The parabolic equation
In this case, $\xi = \eta$. We can choose $Y = \xi = \eta$ and $X = x$ as the independent variables. Then we have

$$\frac{\partial^2 \psi}{\partial X^2} = D'[\psi, \psi_X, \psi_Y, X, Y].$$

The proofs of these are left as exercises (see Exercise 5.13)

5.3.3 The wave equation in two dimensions

Consider the wave equation in two dimensions with $v = 1$: $\left[\frac{\partial^2}{\partial x^2} - \frac{\partial^2}{\partial t^2}\right]\psi = 0$. Let the Cauchy boundary conditions be given on the segment $0 \leq x \leq 1$.

$$\psi(x,0) = \alpha(x) \tag{5.24}$$

$$\frac{\partial \psi}{\partial t}(x,0) = \beta(x) \quad 0 \leq x \leq 1 \tag{5.25}$$

We shall show that this uniquely determines the solution in a domain of the (t, x) plane and determine it.

First, we note that the characteristics given by

$$\xi(x,t) \equiv x - t = C_1$$

$$\eta(x,t) \equiv x + t = C_2 \tag{5.26}$$

are straight lines at 45° with the x-axis so that the open hypersurface $0 \leq x \leq 1, t = 0$ is never parallel to the characteristics at any point on it. Hence, we expect the solution to be determined in some neighbourhood of each point of open hypersurface. We can rewrite the PDE as

$$\frac{\partial^2}{\partial \xi \partial \eta}\psi = 0.$$

The solution to this is immediate:

$$\psi = f_1(\xi) + f_2(\eta).$$

The Cauchy boundary conditions then imply

$$\psi = f_1(x) + f_2(x) = \alpha(x),$$

$$\frac{\partial \psi}{\partial t}(x,0) = -f_1'(x) + f_2'(x) = \beta(x) \quad \text{for} \quad 0 \leq x \leq 1. \tag{5.27}$$

We can integrate the second equation from some arbitrary point $x_0 \in [0, 1]$ as

$$-f_1(x) + f_2(x) = \int_{x_0}^{x} \beta(x')dx' + C, \quad 0 \leq x \leq 1,$$

where C is the constant of integration. We can thus solve for $f_1(x)$ and $f_2(x)$ as

$$f_1(x) = \frac{1}{2}\left\{\alpha(x) - \int_{x_0}^{x} \beta(x')dx' - C\right\},$$

$$f_2(x) = \frac{1}{2}\left\{\alpha(x) + \int_{x_0}^{x} \beta(x')dx' + C\right\}. \tag{5.28}$$

We add the two equations to find that

$$\psi(x,t) = f_1(x-t) + f_2(x+t)$$

$$= \frac{1}{2}\left\{\alpha(x-t) + \alpha(x+t) + \int_{x-t}^{x+t} dx'\beta(x')\right\}. \tag{5.29}$$

The solution is now determined in that region for which $\alpha(x-t)$ and $\alpha(x+t)$ occurring in (5.29) are known. Since $\alpha(x)$ is known for $0 \le x \le 1$ from the boundary conditions (5.25), this region is the intersection of infinite strips defined by

$$0 \le x - t \le 1; \quad \text{and} \quad 0 \le x + t \le 1.$$

This is a finite square area symmetrically situated about the x-axis.

5.4 Some Results Regarding Existence, Uniqueness and Stability of Solutions of a PDE

5.4.1 Results regarding the uniqueness of solutions to the Poisson equation

We shall first consider the Poisson equation in a connected region V bounded by a closed surface S. In the context of future discussions, we may, at times, find it necessary to split the surface S in two open surfaces: $S = S_1 + S_2$. In the following we shall assume the existence of the solution to the Poisson equation and discuss the question of its uniqueness.

We shall show that

(a) (i) There is a unique solution to the Poisson equation within V with a Dirichlet boundary condition on the *closed* surface S, or a mixed boundary condition on the closed surface S.

(ii) There is a unique solution (up to a constant) to the Poisson equation with a Neumann boundary condition on the *closed* surface S[30].

(b) Dirichlet boundary condition (or a Neumann boundary condition) on an *open* surface is *insufficient* to obtain a unique solution.

(c) Cauchy boundary conditions on a closed surface ‘overdetermines’ the problem.

(d) Also, in the context of a 2-dimensional Poisson equation, the Cauchy boundary conditions on an open surface S_1 lead to an unstable solution.

[30]For the Poisson equation, $\nabla^2\phi = -\rho$ in a volume V bounded by the surface S, the Neumann boundary condition must satisfy $\oint_S \frac{\partial\phi}{\partial n} dS = -\int_V dV\rho(\mathbf{r})$. (See Exercise 5.18)

Proof of (a)

Consider the Poisson equation in a region bounded by a surface S with either the Dirichlet, the Neumann or the mixed boundary conditions specified on S. Let, if possible, there be two solutions for ϕ, say ϕ_1 and ϕ_2, for the Poisson equation. Consider $\Phi \equiv \phi_1 - \phi_2$. Then,

(a) Φ satisfies the Laplace's equation on and within S;

(b) $\Phi = 0$ on S in the case of the Dirichlet boundary condition;

(c) $\frac{\partial \Phi}{\partial n} = 0$ on S in the case of the Neumann boundary condition;

(d) $\Phi = 0$ on S_1, and $\frac{\partial \Phi}{\partial n} = 0$ on S_2 in the case of a mixed boundary condition on $S = S_1 + S_2$;

(e) In all cases, $\Phi \frac{\partial \Phi}{\partial n} = 0$ on S.

Now consider the integral,

$$0 = \oint_S \Phi \frac{\partial \Phi}{\partial n} dS = \oint_S \Phi \nabla \Phi . \hat{n} dS \tag{5.30}$$

$$= \int_V \nabla . \{\Phi \nabla \Phi\} dV \tag{5.31}$$

$$= \int_V \{|\nabla \Phi|^2 + \Phi \nabla^2 \Phi\} dV \tag{5.32}$$

$$= \int_V |\nabla \Phi|^2 dV. \tag{5.33}$$

We have used the Gauss' theorem to go from (5.30) to (5.31) and used $\nabla^2 \Phi = 0$ in (5.32). In view of the fact that the last integrand is always nonnegative, the only way the integral over a volume will vanish is to have

$$|\nabla \Phi|^2 = 0 \Rightarrow \nabla \Phi = 0$$

throughout V. This implies that

$$\Phi = \text{constant throughout V}.$$

Thus,

(i) for the Dirichlet boundary condition, $\Phi = 0$ on S. Therefore the constant in the above equation must be zero.

(ii) for a mixed boundary condition we know that $\Phi = 0$ on a part of S. Again, this requires that the constant in the above equation vanishes.

(iii) for the Neumann boundary condition, we know only that $\frac{\partial \Phi}{\partial n} = 0$ on S. This is compatible with any value for the constant in the above equation.

This proves the assertions in (a).

Once we have proven the result (a), it becomes possible to easily demonstrate the results (b) and (c). We shall prove (b) here for the Dirichlet boundary condition, leaving the generalization for the Neumann and the mixed boundary conditions as exercises (see Exercise 5.15).

Proof of (b)

Let S be a closed surface and S_1 be an open section of it. Let S_2 stand for the complementary part of the closed surface S: (i.e., $S = S_1 + S_2$). Let a Dirichlet boundary condition be specified on S_1:

$$\phi|_{S_1} = \phi_1(x, y, z). \tag{5.34}$$

For the sake of argument, let us imagine augmenting the above boundary condition by

$$\phi|_{S_2} = \phi_2(x, y, z). \tag{5.35}$$

The two boundary conditions together constitute a Dirichlet boundary condition on the closed surface S. According to the uniqueness theorem (a), there is a unique solution $\phi = \Phi_1(x, y, z)$ throughout V satisfying this boundary condition. Now, let us consider modifying (5.35) to

$$\phi|_{S_2} = \phi_2'(x, y, z). \tag{5.36}$$

Now (5.34) together with (5.36) also constitutes another Dirichlet boundary condition on a closed surface S. There is, therefore, a unique solution $\phi = \Phi_2(x, y, z)$ throughout V satisfying the second Dirichlet boundary condition. These two solutions cannot be identical, since they differ on S_2. (See equations (5.35) and (5.36).) Thus the two distinct solutions $\phi = \Phi_{1,2}$ both satisfy the PDE and the *same* Dirichlet boundary condition on the *open* surface S_1. This proves non-uniqueness of the solution under these boundary conditions.

In fact, an extension of the above argument will allow one to deduce that there are an infinity of solutions in this case.

Proof of (c)

We shall now prove the result (c) regarding the Cauchy boundary condition on a closed surface. Let it be specified as

$$\phi|_S = \chi_1(x, y, z), \tag{5.37}$$

$$\left.\frac{\partial \phi}{\partial n}\right|_S = \chi_2(x, y, z). \tag{5.38}$$

Let us now consider the PDE together only with the boundary condition of (5.37). This by itself constitutes a Dirichlet boundary condition on a closed surface S. According to the result (a), there is a unique solution to the PDE satisfying this within V. Let this solution be $\phi = \Phi(x, y, z)$. We can now evaluate the derivative $\frac{\partial \Phi}{\partial n}$ on S (from the interior of V). Now, if it so happens by coincidence that the calculated $\frac{\partial \Phi}{\partial n}$ on S agrees with that given in (5.38), then (and only then) we have a solution to the PDE with this (Cauchy) boundary condition. In this case, either (5.37) *or* (5.38) is superfluous. In any other case, there is *no solution* to the PDE with the present boundary condition.

In this case, we say that the boundary condition 'overdetermines the problem' in the sense that one is specifying too much of boundary conditions than is required for a unique solution to exist. In such a case, barring coincidence, there will be no solution to the PDE.

Proof of (d)

We shall finally deal with the Cauchy boundary conditions on an open surface. Here, we shall find it convenient to specialise to two dimensions. We wish to show the instability of the solution for this kind of a boundary condition. We do this by first considering a particular example and then clarifying the general reasoning.

Consider the Laplace's equation in two dimensions in the *interior* of a square together with the Cauchy boundary conditions as give below.

$$\left(\frac{\partial^2}{\partial x^2} + \frac{\partial^2}{\partial y^2}\right)\Phi(x, y) = 0 \quad 0 < x < 1; \quad 0 < y < 1 \tag{5.39}$$

Consider two of the following solutions $\Phi_1(x, y)$ and $\Phi_2(x, y)$ to it and the Cauchy boundary conditions they satisfy on $0 < x < 1$.

1. $$\Phi_1(x, y) \equiv 0 \quad 0 < x < 1; \quad 0 < y < 1$$

   which satisfies the Cauchy boundary conditions on an open hypersurface $0 < x < 1$:

   $$\Phi_1(x, 0) = 0 \qquad 0 < x < 1$$

   $$\left.\frac{\partial \Phi_1(x, y)}{\partial y}\right|_{y=0} = 0 \qquad 0 < x < 1$$

2. $$\Phi_2(x, y) = \epsilon \operatorname{Re} \frac{1}{z - \frac{1}{2} - i} = \frac{\epsilon(x - 1/2)}{\left[\left(x - \frac{1}{2}\right)^2 + (y-1)^2\right]}$$

   $$0 < x < 1; \quad 0 < y < 1 \tag{5.40}$$

In the above, ϵ is a arbitrary positive parameter which we can choose as small as we wish. Φ_2 satisfies the Laplace's equation, being real part of a function analytic in the interior of the square. It satisfies the Cauchy boundary conditions on the same open hypersurface:

$$\Phi_2(x,0) = \frac{\epsilon(x-1/2)}{[(x-1/2)^2+1]} \qquad 0 < x < 1$$

$$\left.\frac{\partial\Phi_2(x,y)}{\partial y}\right|_{y=0} = \frac{2\epsilon(x-1/2)}{[(x-1/2)^2+1]^2} \qquad 0 < x < 1$$

We note that on the open hypersurface $0 < x < 1$ the two boundary conditions differ by

$$|\Phi_2(x,0) - \Phi_1(x,0)| = \frac{\epsilon\,|x-1/2|}{[(x-1/2)^2+1]} < \epsilon/2,$$

$$\left|\frac{\partial[\Phi_2(x,y) - \Phi_1(x,y)]}{\partial y}\right|_{y=0} = \left|\frac{2\epsilon(x-1/2)}{[(x-1/2)^2+1]^2}\right| < \epsilon. \tag{5.41}$$

Thus, these Cauchy boundary conditions differ by a 'small' amount $\sim \epsilon$. On the other hand, consider the difference of the solutions

$$|\Phi_2 - \Phi_1| = \frac{\epsilon|x-1/2|}{[(x-1/2)^2+(y-1)^2]}.$$

This can become arbitrarily large (for any fixed $\epsilon > 0$) as one approaches $(x, y) \to (1/2, 1)$ from within the square. (For example, at $x - 1/2 = \epsilon = 1 - y$, we find that the difference is 1/2!) Thus, a small difference $[O(\epsilon)]$ in the Cauchy boundary conditions can be accompanied by a large difference $[O(1)]$ in the solution in some region! In this case, we say that the solution is *unstable* and hence physically unacceptable.

Now, we can argue the same result also for the case of the Poisson equation in two dimensions. We can add the same *particular* solution of the Poisson equation to both Φ_1 and Φ_2 and thus arrive at two solutions of the *Poisson* equation. These likewise differ by small amounts on the boundary, yet differ significantly in some region.

We shall briefly explain the reasoning behind this result. Consider the Laplace's equation in a domain D of the x–y plane. Any solution Φ_1 of it can be written as the real part of some analytic function $f(z)$.

$$\Phi(x,y) = \text{Re}\, f(z), \quad z \in D.$$

Now consider $g(z)$, another function analytic in D, but with a singularity at a point P just outside the boundary C of D. Let $g(z)$ have no other singularities in the vicinity of C. Let C_1 be an open segment of C that excludes the points in the vicinity of P. Then the function

$$\Phi_2(x, y) \equiv \text{Re}\,[f(z) + \epsilon g(z)],$$

where ϵ is a small number, differs from $\Phi_1(x, y)$ by a 'small' amount of $O(\epsilon)$ on C_1 but approaches $\pm\infty$ as one approaches the point P from the interior of D.

More general examples of this type can easily be constructed. A further example is found as an exercise (see Exercise 5.17).

One could give a similar 'physical' construction in three dimensions by identifying the function $\Phi(\mathbf{r})$ with the electrostatic potential. The example below (see part (e)) illustrates such a construction.

Solved problem 5.4 Consider the Laplace's equation $\nabla^2\psi = 0$ in the region V between the spherical shells[31] $r = R_1$ and $r = R_2 > R_1$.

(a) Find the solution for ψ in V given $\psi|_{r=R_1} = C_1$ and $\psi|_{r=R_2} = C_2$.

(b) Find under what conditions is there a solution for ψ in V given $\left.\frac{\partial\psi}{\partial n}\right|_{r=R_1} = D_1$ and $\left.\frac{\partial\psi}{\partial n}\right|_{r=R_2} = D_2$. Explain the reasons.

(c) Find the solution for ψ in V given $\left.\frac{\partial\psi}{\partial n}\right|_{r=R_1} = D_1$ and $\psi|_{r=R_2} = C_2$.

(d) Write down a class of solutions satisfying the Dirichlet boundary condition on the *open* surface: $\psi|_{r=R_1} = C_1$.

(e) Show by constructing an example that the solution with the Cauchy boundary conditions $\psi|_{r=R_1} = C_1$; $\left.\frac{\partial\psi}{\partial n}\right|_{r=R_1} = D_1$ on the *open* hypersurface is *unstable*.

Solution: The cases (a), (b) and (c) are ones with the Dirichlet, Neumann and mixed boundary conditions on a closed, albeit multiply connected surface. Given a solution, the theorems proved earlier guarantee its uniqueness. Thus, a solution can be obtained by guessing its form and verifying the boundary conditions. Here, the spherical symmetry of the problem is useful in guessing the result.

(a) We try the solution $\psi = \frac{A}{r} + B$. We see that this ψ satisfies the Laplace's equation in V. Further, it satisfies the boundary conditions if $B = \frac{R_2C_2 - R_1C_1}{R_2 - R_1}$, $A = \frac{(C_1 - C_2)R_1R_2}{R_2 - R_1}$.

(b) Again, we expect a spherically symmetric solution and attempt a solution of the form $\psi = \frac{A}{r} + B$. We find that the normal derivatives along the *outward* normal are given by

[31] In this case, the volume V is not a simply-connected region and its boundary has two disconnected pieces. Nonetheless, $S_1 + S_2$ forms a closed boundary of V.

$$\left.\frac{\partial\psi}{\partial n}\right|_{r=R_1} = -\left.\frac{\partial\psi}{\partial r}\right|_{r=R_1} = \frac{A}{R_1^2} = D_1,$$

$$\left.\frac{\partial\psi}{\partial n}\right|_{r=R_2} = \left.\frac{\partial\psi}{\partial r}\right|_{r=R_2} = -\frac{A}{R_2^2} = D_2.$$

Thus, a solution exists only if $D_1R_1^2 + D_2R_2^2 = 0$. This, of course, is the condition that $\oint_S dS\frac{\partial\psi}{\partial n} = 0$ for a Laplace's equation. Given this, we find that the PDE and the boundary conditions are satisfied if

$$A = D_1R_1^2, \text{ and B is arbitrary.}$$

(c) In this case we similarly find that $\psi = \frac{A}{r} + B$ with $A = R_1^2D_1$, $B = C_2 - \frac{R_1^2D_1}{R_2}$ is the solution.

(d) We note that the one parameter family of solutions $\psi = \frac{A}{r} + B$ with $B = \frac{R_2C_2-R_1C_1}{R_2-R_1}$, $A = \frac{(C_1-C_2)R_1R_2}{R_2-R_1}$, where C_2 is any arbitrary real number, is a class of solutions to the Laplace's equation with the Dirichlet boundary condition on an open hypersurface $r = R_1$.

(e) Let Ψ_1 be one solution to the given boundary conditions. Consider another solution Ψ_2 defined by

$$\Psi_2 = \Psi_1 + \frac{\epsilon}{|\mathbf{r} - (R_2+\delta)\hat{i}|},$$

where ϵ and δ are small positive members. We note that $\Psi_2 - \Psi_1$ is proportional to the potential of a small positive charge placed at $x = R_2 + \delta$, $y = 0$, $z = 0$, i.e., just outside the shell. It is not difficult to verify that the Cauchy boundary conditions on the sphere of radius R_1 for the two solutions differ by $O\left(\frac{\epsilon}{R_2-R_1}\right)$ for ψ and (utmost) by $O\left(\frac{\epsilon}{(R_2-R_1)^2}\right)$ for the normal derivative. On the other hand the two solutions differ by a quantity of $O\left(\frac{\epsilon}{\delta}\right)$ near the point $x = R_2$, $y = 0$, $z = 0$ and this can be made arbitrarily large by taking δ arbitrarily small with ϵ fixed.

5.4.2 Stability of the solution to Poisson equation with the Dirichlet boundary condition on a closed surface

In this subsection, we shall give a brief argument that enables one to understand the stability of the solution to the Poisson equation with a Dirichlet boundary condition on a closed hypersurface.

First, we shall note a generalization of the result regarding the maxima and minima of the solution of the Laplace's equation which we have noted earlier for the 2-dimensional case.

We shall consider an n-dimensional Euclidean space. Let Φ satisfy the Laplace's equation in n dimensions: $\sum_{i=1}^{n} \frac{\partial^2}{\partial x_i^2}\Phi = 0$ in a volume V. We shall then show that Φ attains its maxima and mimina on the boundary of V. Let, if possible, P be a point in the interior of V where Φ has its (local) maximum. This would then imply that the matrix of second derivatives at P, viz., $A_{ij} = \left[\frac{\partial^2}{\partial x_i \partial x_j}\Phi\right]$ has all eigenvalues negative. Then $\mathrm{tr}[A] = \frac{\partial^2}{\partial x_i^2}\Phi$ must then be negative at P. But this is impossible since Φ satisfy the Laplace's equation there. The same argument goes through for a (local) minimum. Hence, etc.

We shall use the above argument to demonstrate the stability of the solution to the Poisson equation in a (simply-connected) volume V, given the Dirichlet boundary condition on its surface S. Let ϕ_1 be a solution of the Poisson equation in $\{V + S\}$. Consider perturbing the boundary condition on S by a small amount. Let ϕ_2 be the solution associated with this perturbed boundary condition. Then $|\phi_2 - \phi_1|$ is small on S. Let, to be concrete, $|\phi_2 - \phi_1|_S < \epsilon$. Then it is easy to see that $|\phi_2 - \phi_1| < \epsilon$ throughout V. This is so since $\Phi \equiv \phi_2 - \phi_1$ satisfies the Laplace's equation in V and hence $|\Phi|$ has its maximum on the boundary where it is less than ϵ.

5.4.3 Results regarding the hyperbolic equation

We shall restrict the discussion to the simplest hyperbolic equation in two dimensions – the wave equation. It reads (we set $c = 1$ for simplicity)

$$\frac{\partial^2 \psi}{\partial t^2} - \frac{\partial^2 \psi}{\partial x^2} = 0. \tag{5.42}$$

We shall mainly confine ourselves to illustrating some results about the various boundary conditions for this equation with examples. These are:

(a) There is a unique solution to the PDE with a Cauchy boundary condition on a open curve, in a region determined by the curve, provided the curve nowhere is parallel to the characteristics.

(b) Cauchy boundary conditions on a closed hypersurface overdetermine[32] the solution.

(c) Dirichlet boundary conditions on an *open* hypersurface are insufficient to determine a unique solution. The same is valid for the Neumann boundary conditions OR the mixed boundary conditions on an open hypersurface.

(d) Dirichlet boundary conditions on a closed hypersurface may not lead to a unique solution.

As stated earlier, we can give only illustrating examples or arguments for these statements.

---

[32] By 'overdetermining' a solution, we mean specifying more information than is necessary for a unique solution.

(a) In Section 5.3.3, we have already seen that there is an analytic solution in some neighbourhood of each point of the curve C under the conditions in (a) above. We have in fact shown that the solution is determined in a square in the (x, t) plane. We shall exploit this example to illustrate the succeeding statements.

(b) Consider the Cauchy boundary conditions for the wave equation on a rectangle with sides parallel to axes given by equations: (i) $t = 0$, (ii) $t = 1/4$, (iii) $x = 0$, (iv) $x = 1$. Now the Cauchy boundary conditions on $0 \le x \le 1$ at $t = 0$ already determine a unique solution in the square that contains the line segment $t = 1/4$, $1/4 \le x \le 3/4$. Thus, there is no further freedom to specify any boundary conditions on this segment. Thus, specifying the Cauchy boundary conditions on this segment would overdetermine the solution. For another example, please see Exercise 5.6.

(c) Suppose that the Dirichlet boundary condition is specified on an open curve C. This can be augmented by two distinct Neumann boundary conditions on the same segment. These then constitute two *distinct* Cauchy boundary conditions on C. According to (a), each leads to a unique but distinct solution in the vicinity of C. Thus the Dirichlet boundary condition on an open curve C does not have a unique solution associated with it.

(d) An example of this has been seen in Section 5.1. (See part (b) of solved problem 5.1.)

Exercises

5.1 Consider the most general linear elliptic equation in 3 dimensions in the Cartesian coordinates.

$$\sum_{i,j=1}^{3} a_{ij}(x_1, x_2, x_3)\frac{\partial^2 \Phi}{\partial x_i \partial x_j} + \sum_{i=1}^{3} b_i(x_1, x_2, x_3)\frac{\partial \Phi}{\partial x_i} + c(x_1, x_2, x_3)\Phi = d(x_1, x_2, x_3)$$

Show that there exists a coordinate rotation that can put it in the diagonal form at a given point

$$\sum_{i,j=1}^{3} \lambda_i(x_1', x_2', x_3')\frac{\partial^2 \Phi}{\partial x_i'^2} + \sum_{i=1}^{3} b_i'(x_1', x_2', x_3')\frac{\partial \Phi}{\partial x_i'} + c'(x_1', x_2', x_3')\Phi = d'(x_1', x_2', x_3')$$

with all $\lambda_i > 0$. Show further that this equation is invariant in form under all rotations and translations only if

$$\lambda_1 = \lambda_2 = \lambda_3 = \lambda; \quad b_i \equiv 0; \quad c = \text{const}; \quad d = \text{const}.$$

Show that in this case, the equation reduces to either (i) Poisson equation ($c = 0$) (with a constant source d), or (ii) Helmholtz wave equation with a constant 'source' d, if $c > 0$. The Poisson equation and the Helmholtz wave equation are the two elliptic equations that occur frequently in physics.

5.2 In a manner similar to the above problem, show that the only linear homogeneous second order hyperbolic type PDE invariant under Poincare transformations (which contain the Lorentz transformations and space–time translations) is the 'Klein–Gordon' equation: $\left[\nabla^2 - \frac{1}{c^2}\frac{\partial^2}{\partial t^2} - m^2\right]\psi = 0$.

5.3 State the type of the following differential equation and obtain the equations for its characteristics.

$$\frac{\partial^2 u}{\partial x^2} - y^2\frac{\partial^2 u}{\partial y^2} - y\frac{\partial u}{\partial y} = 0$$

Sketch a few of the characteristics. Given the Cauchy boundary condition that

(a) $u(x, y_0) = a(x)$

(b) $\left.\frac{\partial u}{\partial y}\right|_{y=y_0} = b(x)$

with $y_0 > 0$, show that the solution for $y > y_0$ is

$$u = \frac{1}{2}\left[a(\alpha) + a(\beta) + y_0\int_\beta^\alpha b(x')dx'\right],$$

where

$$\alpha = x + \ell n\frac{y}{y_0}; \quad \beta = x - \ell n\frac{y}{y_0}.$$

5.4 Consider the wave equation ($c = 1$)

$$\frac{\partial^2 u}{\partial x^2} - \frac{\partial^2 u}{\partial t^2} = 0$$

with the boundary conditions:

$$u(x, 0) = a(x)$$

$$\left.\frac{\partial u}{\partial t}\right|_{t=0} = b(x)$$

for $0 \le x \le a$; and

$$u(0, t) = 0 = u(a, t) \quad \text{for} \quad 0 \le t \le a.$$

Determine the solution and show the area in the plane in which the solution is determined.

Consider, now, changing the last boundary condition to

$$u(0, t) = 0 = u(a, t) \quad \text{for} \quad 0 \leq t \leq na \quad n = 2, 3, \ldots.$$

Develop an argument to show that by letting $n \to \infty$ one can determine $u(x, t)$ in the infinite strip bounded by $t = 0$, $x = 0$ and $x = a$.

5.5 Repeat the first part of the previous problem with the second boundary condition

$$\left.\frac{\partial u}{\partial x}\right|_{x=0} = \left.\frac{\partial u}{\partial x}\right|_{x=a} = 0; \quad 0 \leq t \leq a;$$

(i.e., the Neumann boundary condition). Consider also the other combinations of Dirichlet and Neumann boundary condition at $x = 0$ and $x = a$.

5.6 Consider the hyperbolic equation

$$\frac{\partial^2 u}{\partial x^2} - \frac{\partial^2 u}{\partial y^2} = 0$$

and a closed boundary that consists of a square of side a in the first quadrant with one corner at the origin. Show that C nowhere coincides with any of the characteristics. Show that giving Cauchy conditions over $0 \leq x \leq a$ and Dirichlet or Neumann conditions over the rest of C overdetermines the solution.

5.7 Determine the solution to the Laplace's equation in two dimensions:

$$\frac{\partial^2 u}{\partial x^2} + \frac{\partial^2 u}{\partial y^2} = 0$$

with the Dirichlet boundary condition that

$$u|_{r=a} = A(\theta); \quad r = \sqrt{x^2 + y^2}.$$

Repeat the exercise the Neumann boundary condition, viz.,

$$\left.\frac{\partial u}{\partial r}\right|_{r=a} = B(\theta).$$

What would happen if Cauchy conditions were given on the circle $r = a$?

5.8 Consider the Laplace's equation in two dimensions: $\left[\frac{\partial^2}{\partial x^2} + \frac{\partial^2}{\partial y^2}\right]\phi = 0$ with the Dirichlet boundary conditions on a square of sides L:

(i) $\phi = 1$ for $0 < y < L$ and $x = 0$;
(ii) $\phi = 3$ for $0 < y < L$ and $x = L$;

(iii) $\phi = 2$ for $0 < x < L$ and $y = L$;
(iv) $\phi = 4$ for $0 < x < L$ and $y = 0$.

Place upper and lower bounds on $\phi(x, y)$ for $0 < x, y < L$. [Hint: Recall the result regarding the maxima and minima of a solution to the Laplace's equation.]

5.9 Solve the Laplace's equation in two dimensions: $\left[\frac{\partial^2}{\partial x^2} + \frac{\partial^2}{\partial y^2}\right]\phi = 0$ by the method of separation of variables and using the boundary conditions on a square: $\phi = 0$ for $0 < y < L$ and $x = 0$; and $\phi = 0$ for $0 < y < L$ and $x = L$. Discuss the uniqueness of the solution with each of the following additional boundary conditions:

(i) $\phi = 2$ for $0 < x < L$ and $y = L$ and $\phi = 4$ for $0 < x < L$ and $y = 0$;
(ii) $\frac{\partial \phi}{\partial x} = 0$ for $0 < x < L$ and $y = L$ and $\frac{\partial \phi}{\partial x} = 0$ for $0 < x < L$ and $y = 0$;
(iii) $\frac{\partial \phi}{\partial y} = 0$ for $0 < y < L$ and $x = L$ and $\frac{\partial \phi}{\partial y} = 0$ for $0 < y < L$ and $x = 0$.

5.10 Show that given $\Delta \neq 0$ in some neighbourhood of a point (x_0, y_0) on C, all-order derivatives of $\psi(x, y)$ can be determined there.

5.11 Consider the Laplace's equation (for two dimensions) within a unit square in the first quadrant with edges parallel to axes and one corner at (0, 0).

(a) Show that $\Phi(x, y) = \sin(kx)\sinh(ky) \qquad 0 < x, y < 1$
(k is any real constant) is a solution to it.
(b) State (an example each of) an explicit boundary condition for which this $\Phi(x, y)$ will be (i) a unique stable solution within the square, (ii) an unstable solution within the square, (iii) a solution within the square, but not a unique one.

5.12 Consider the differential equations:

(a) $\left[\frac{\partial^2}{\partial x^2} + \frac{\partial^2}{\partial y^2}\right]\psi = 0$

(b) $\left[\frac{\partial^2}{\partial x^2} - \frac{\partial^2}{\partial y^2}\right]\psi = 0$

(c) $\left[\frac{\partial^2}{\partial x^2} - \frac{\partial}{\partial y}\right]\psi - b = 0 \qquad b \neq 0$

within a square of side a.
Consider the boundary condition: $\psi(x, 0) = 0; \quad 0 \le x \le a$. Show that, with this boundary condition the PDE's (a) and (b) have an infinity of solutions $\psi(x, y) = cy$; $c =$ an arbitrary constant while the PDE (c) has a unique solution $\psi(x, y) = by$.

5.13 Prove the standard form of the PDE's in two dimensions with constant coefficients A, B, C given in Section 5.3.2

5.14 Show that the answers one gets from (5.13) and (5.14) are identical.

5.15 Generalize the result (b) in the Section 5.4.1, to Neumann and mixed boundary conditions.

5.16 Consider an n-dimensional hypersurface (either closed or open) on which the tangential derivatives of y are given at each point. Show then that y is known on the surface up to a constant.

5.17 Consider the two solutions of the Laplace's equation in the x–y plane:

$$\Phi_1(x, y) = 0$$

$$\Phi_2(x, y) = \frac{1}{k^2} \sin(kx) \sinh(ky)$$

Show that for sufficiently large k, these two solutions satisfy Cauchy boundary conditions on the x-axis that are arbitrarily close, while the solutions themselves can differ by a large number in some regions of the x–y plane.

5.18 For the Poisson equation $\nabla^2 \phi = -\rho$ in a volume V bounded by the surface S, show that the Neumann boundary condition cannot be specified arbitrarily, but must satisfy: $\oint_S \frac{\partial \phi}{\partial n} dS = -\int_V dV \rho(\mathbf{r})$. [Hint: Integrate the Poisson equation over V and use the Gauss' theorem.]

6 Generalized Functions, the Dirac Delta Function and the Fourier Transform

6.1 Introduction

The Dirac delta function is ubiquitous in theoretical physics. Students may encounter it in quantum mechanics, in electrodynamics and in general in any branch of advanced theoretical physics. To give an example, consider the following simple illustration. As we know, we always use various idealizations in physics as these are mathematically simpler to deal with, one of them being that of the point charge. If one asks what function describes the charge density of a point charge q at $r = 0$, it is clearly a 'function' with the properties

$$\rho_P(\mathbf{r}) = 0 \quad r \neq 0$$

$$\rho_P(\mathbf{r}) \to \infty \quad r \to 0$$

such that $\int \rho_P(\mathbf{r}) d^3r = q$. Clearly, $\rho_P(\mathbf{r})$ is not a mathematical 'function' in the sense that it is not well defined at the origin where the point charge is located. We cannot make it well defined by omitting the singular point $r = 0$ (as we often do for mathematical functions) because then we would miss the whole physical context – the charge which gave rise to a field. In spite of the fact that $\rho_P(\mathbf{r})$ is not a mathematically well-defined function, we do use the idea of a point charge at the very basis of calculations in electrostatics. This is also true of the (volume) charge density associated with a line charge, a surface charge, and so on.

The purpose of this chapter is to give a mathematically rigourous formalism of 'generalized functions' that, in particular, will place on a sound footing the 'layman's' usage of singular functions such as these. It will also enable us to treat the Green's functions (whose definitions involve the Dirac δ-function) on a rigourous footing. Arbitrary differentiations under the integral sign (in say the Fourier transforms) require for their rigour the use of generalized functions. In particular, the use of Fourier transform method in the solutions of differential equations requires for its rigour the formalism of generalized functions.

It is perhaps worth considering how and why we can use these idealizations in computations. A 'point charge' in a macroscopic problem in electrostatics is in actuality a charge concentrated over such a small region in space as compared to the dimensions of the problem. The point charge is a mathematically convenient limit that allows two things: (i) it simplifies algebra involved in computations and at the same time, (ii) it allows for evaluation of physical quantities to a high level of accuracy. We need not, in fact, use the notion of a concentrated finite point charge as such; we could do all the calculations with q distributed over a small sphere which then would do away with singular functions. But then we would have to involve ourselves into unnecessary algebraic complications without any essentially new insights into the physics involved.

The above procedure of distributing q over a sphere then is to regard the charge density $\rho(\mathbf{r})$ as the function

$$\rho_\epsilon(\mathbf{r}) = 0 \qquad r > \epsilon,$$

$$\rho_\epsilon(\mathbf{r}) = \frac{q}{\frac{4}{3}\pi\epsilon^3} \qquad r < \epsilon.$$

The above function $\rho_\epsilon(\mathbf{r})$ describes the charge density of a small uniformly charged sphere of radius ϵ. $\rho(\mathbf{r})$ is finite everywhere and is thus a well-defined mathematical function. We now regard a point charge q as given by the limit

$$\rho_P(\mathbf{r}) = \lim_{\epsilon\to 0} \rho_\epsilon(\mathbf{r}). \tag{6.1}$$

We can thus regard the singular 'function' $\rho_P(\mathbf{r})$ as the limit of a sequence of functions $\{\rho_{\epsilon/n}(\mathbf{r}); \epsilon > 0 \; n = 1, 2, 3, \ldots\}$ parameterised by a parameter n, each of which is a mathematically well-defined function. We can look upon the charge density of a point charge in terms of such a sequence.

We further note that (6.1) is not the only way we can regard the charge density of point charge. We could define

$$\rho_\sigma(\mathbf{r}) = \frac{1}{(\sqrt{2\pi}\sigma)^3} e^{-r^2/2\sigma^2} \qquad \sigma > 0$$

and consider

$$\rho_P(\mathbf{r}) = \lim_{\sigma\to 0} \rho_\sigma(\mathbf{r}).$$

We thus see that $\rho_P(\mathbf{r})$ has been now seen as the limit of a sequence of ordinary (well-defined) functions $\{\rho_{\sigma/n}(\mathbf{r}) : n = 1, 2, 3, \ldots\}$. (This latter class of functions is moreover *infinitely differentiable* unlike the former and hence more tractable in calculations.)

We do know from the experience in electrostatics that the results for a point charge do not depend on how we define $\rho_P(\mathbf{r})$.

We shall now summarise the above discussion and raise a few pertinent questions.

1. Many idealizations in physics are represented by functions which are not mathematically well-defined.
2. For our convenience at least, we cannot do away with these ill-defined functions in practice.
3. Such an idealization as a point charge is understood in a mathematically rigourous fashion by interpreting it as a 'limit of a sequence of well-defined functions'. We shall relate such sequences (appropriately chosen) to the nation of generalised functions.
4. A generalised function will be understood (as we shall see soon) as a sequence of an infinite set of well-defined functions (with certain necessary properties to be defined below).
5. For a point charge, for example, there may be a number of distinct sequences whose limit gives $\rho_p(\mathbf{r})$.
6. Certain mathematical results may be insensitive to how $\rho_p(\mathbf{r})$ is defined. (One needs to formulate when this happens.)

As emphasised earlier, we could do a typical calculations for point charges involved in electrostatics by using uniformly charged spheres of radius ϵ and take the limit $\epsilon \to 0$ in the calculations in the end. This procedure would be mathematically rigourous though cumbersome.

Our aim in the section is to develop directly the calculus of the so-called generalised functions so that *this cumbersome procedure can be avoided without sacrificing rigour.* We would also like to know the mathematical conditions under which this cumbersome procedure can be avoided. These are evolved in a general manner in this chapter.

The discussion in this chapter will, among other things, help a physics student to place the use of Dirac δ-function on a sound mathematical foundation.

To close the introduction, let us recall several rudimentary situations where a new student of advanced physics encounters δ-function. The aim here is to give a representative sample.

(i) Eigenfunction of a position operator in quantum mechanics: $\delta^3(\mathbf{r} - \mathbf{r}_0)$.

(ii) Charge density of an electron (thought to be a point particle so far as we know today) $= e\delta^3(\mathbf{r} - \mathbf{r}_0)$.

(iii) Closure relation in quantum mechanics

$$\sum_n \psi_n(\mathbf{r})\psi_n^\star(\mathbf{r}') = \delta^3(\mathbf{r} - \mathbf{r}'),$$

where $\{\psi_n(\mathbf{r})\}$ are a complete set of orthonormal wave functions for any physical system.

(iv) Equations determining Green's functions
For electrostatics:

$$\nabla_r^2 G(\mathbf{r}-\mathbf{r}') = -\frac{1}{\epsilon_0}\delta^3(\mathbf{r}-\mathbf{r}')$$

For Schrödinger equation:

$$\left(i\hbar\frac{\partial}{\partial t} - H\right) G((\mathbf{r},t;\mathbf{r}',t') = \delta(t-t')\delta^{(3)}(\mathbf{r}-\mathbf{r}')$$

For the wave equation

$$\left\{\frac{\partial^2}{c^2\partial t^2} - \nabla^2\right\} G((\mathbf{r},t;\mathbf{r}',t') = \delta(t-t')\delta^{(3)}(\mathbf{r}-\mathbf{r}')$$

(v) Solution to the Schrödinger equation with a δ-function potential is used as a good approximation in theoretical treatments in condensed matter physics and nuclear physics.

We note in passing that the above equations defining the Green's functions themselves contain the Dirac delta function. Rigourous treatment of the above equations requires that we look upon the Green's functions as generalised functions.

6.2 Generalised Functions

6.2.1 Some definitions

We would first like to introduce[33] certain classes of functions. These functions have been defined keeping in mind the desirability of certain properties that depend on their differentiability and on their asymptotic behaviour at $\pm\infty$. While giving their definitions, we shall consider functions (possibly complex) of a single real-variable x over the range $-\infty < x < \infty$.

Definition:

A good function A function $f(x)$ is called a good function if the following conditions are fulfilled.

(1) The function and all its derivatives $f^{(n)}(x)$; $n = 0, 1, 2, \ldots$ [with $f^{(0)}(x) = f(x)$] exist everywhere.

(2) $|f^{(n)}(x)|$ vanishes faster than any power of $\frac{1}{|x|}$ as $|x| \to \infty$. In other words, given any $n \geq 0$ and $N > 0$, it is possible to write

[33] See e.g., *Introduction to the Fourier Analysis and Generalised Functions* by M.J. Lighthill, Cambridge Press.

$$|f^{(n)}(x)| < \frac{A}{|x|^N} \quad \text{for} \quad |x| > L \tag{6.2}$$

with some $A > 0$ and L. (Here A and L may depend on n and N).

As the examples below show, the class of good functions is a very restrictive class having 'infinite differentiability' and a highly restrictive asymptotic behaviour. The reason for restricting to this class will become clear in Section 6.5.

We shall also need a somewhat less restrictive class of what are called 'fairly good functions' that includes good functions as a subclass.

Definition: A function $f(x)$ is called a 'fairly good function' if

(1) the function and all its derivatives $f^{(n)}(x) n = 0, 1, 2, \ldots$ exist everywhere;

(2) $|f^{(n)}(x)|$ for each $n = 0, 1, 2, \ldots$ is bounded by some power $P(n)$ of $|x|$ as $|x| \to \infty$. In other words, given $n \geq 0$, there exist positive numbers $A(n)$ and $L(n)$ such that we can write

$$|f^{(n)}(x)| < A(n)|x|^{P(n)} \quad \text{for} \quad |x| > L(n) \tag{6.3}$$

for some $P(n)$.

We note:

1. $P(n)$ for a given n need not have a fixed sign (positive or negative) and can be n-dependent.
2. The class of fairly good functions is distinguished from the class of good functions by the weaker requirement on their asymptotic behaviour.
3. Each good function is, in particular, a fairly good function.
4. If a relation of the form (6.3) is true for some $P(n)$, it is also trivially true for a $\tilde{P}(n) > P(n)$.

Examples of good functions

The following are some examples of 'good functions'.

$$g_1(x) = e^{-x^2}$$

$$g_2(x) = x^n e^{-x^2 A + Bx} \quad A > 0, \quad -\infty < B < \infty.$$

$$g_3(x) = P(x) e^{-Ax^2} \quad A > 0, \quad \text{and } P(x) \text{ a polynomial of an arbitrary order.}$$

$$g_4(x) = \sin x e^{-x^4 A} \quad A > 0$$

$$g_5(x) = 0 \quad x \leq 0$$
$$= x e^{-1/x} e^{-x^2} \quad x > 0$$

On the other hand, the following are examples of functions that are not good functions (reasons indicated in brackets).

$b_1(x) = e^{-x}$ (not bounded as $x \to -\infty$)

$b_2(x) = e^{-|x|}$ ($b_2'(0)$ does not exist)

$b_3(x) = 0 \quad x \le 0$

$= (ax + bx^2)e^{-x} \quad x > 0$ with $b > a > 0$ ($b_3'(0)$ and $b_3''(0)$ do not exist)

$b_4(x) = 0 \quad |x| \ge a$

$= \frac{1}{a} \quad |x| < a$ ($b_4'(\pm a)$ does not exist)

$b_5(x) = \sqrt{|x|}e^{-x^2}$ ($b_5'(0)$ does not exist)

$b_6(x) = \frac{1}{x}e^{-x^2}$ ($b_6(0)$ does not exist)

Examples of fairly good functions

$f_1(x) = g_1(x)$, a good function

$f_2(x) = P(x)$, a polynomial

$f_3(x) = \frac{1}{1+x^2}$

$f_4(x) = \frac{P(x)}{Q(x)}$

where $P(x)$ and $Q(x)$ are polynomials of arbitrary orders with $Q(x) \neq 0$ for $-\infty < x < \infty$.

$f_5(x) = \sin x$

$f_6(x) = \frac{\sin x}{x} \quad x \neq 0$

$= 1 \quad x = 0$

are examples of fairly good functions whereas the following are not examples of fairly good functions for reasons indicated in brackets.

$b_1(x) = |x|$ ($b_1'(0)$ does not exist)

$b_2(x) = \sin|x|$ ($b_1'(0)$ does not exist)

$b_3(x) = \sqrt{x}$ ($b_1'(0)$ does not exist)

$b_4(x) = \frac{\sin x}{x^2}$ ($b_4(0)$ does not exist)

$$b_5(x) = e^x \quad \text{(cannot be bounded by a power as } x \to \infty)$$

$$b_6(x) = x^n \ln|x| \quad (b_6^{(n+1)}(0) \text{ does not exist})$$

Solved problem 6.1 Verify directly from the definition that

(a) $g(x) = e^{-x^2}$ is a good function.
(b) $b(x) = |x|e^{-x^2}$ is not a good function.
(c) $f(x) = x^3$ is a fairly good function.
(d) $b(x) = |x|^3$ is not a fairly good function.

Solution:

(a) We note:

(i) $g(x)$ is infinitely differentiable.
(ii) $g^{(n)}(x)$ has the form $P_n(x)e^{-x^2}$ where $P_n(x)$ is some polynomial in x. (Prove!)
(iii) $x^m e^{-x^2} \to 0$ as $|x| \to \infty$ for any m.

Hence $g^{(n)}(x) \to 0$ faster than any power of $\frac{1}{|x|}$ as $|x| \to \infty$ for $n = 0, 1, 2, \ldots$. Thus $g(x)$ is a good function.

(b)
$$\begin{aligned} b(x) &= |x|e^{-x^2} \\ &= xe^{-x^2} \quad x > 0 \\ &= -xe^{-x^2} \quad x \leq 0 \end{aligned}$$

$b(x)$ is a product of two functions $|x|$ and e^{-x^2}, of which ($|x|$) is not differentiable at $x = 0$. Hence, we carry out the test for $b'(0)$. We have,

$$b'(x) = e^{-x^2} - 2x^2e^{-x^2} \quad x > 0,$$
$$b'(x) = -e^{-x^2} + 2x^2e^{-x^2} \quad x < 0.$$

Thus, $\lim_{x\to 0^+} b'(x) = 1$; $\lim_{x\to 0^-} b'(x) = -1$.
Hence, $b'(0)$ does not exist. Hence $b(x)$ is not a good function.

(c) $b(x) = x^3$ is infinitely differentiable with

$$b'(x) = 3x^2 \quad b''(x) = 6x \quad b'''(x) = 6; \quad b^{(n)}(x) = 0, \quad n > 3.$$

Each of these is clearly power-bounded. For completeness, we can write the bounds of the form (6.3):

$$b'(x) < 4|x|^2; \quad |b''(x)| < 7|x|; \quad |b'''(x) < 7.$$

We can also write collectively

$$|b^{(n)}(x)| < 7|x^3| \quad |x| > 1 \quad \text{for all} \quad n.$$

(d) $\tilde{b}(x) = |x|^3$ is not infinitely differentiable as

$$\begin{aligned} \tilde{b}'''(x) &= +6 \quad x > 0 \\ &= -6 \quad x < 0 \end{aligned}$$

and $\tilde{b}'''(0)$ does not exist. Hence $\tilde{b}(x)$ is a not a fairly good function.

Solved problem 6.2 Show that the function $S(x)$ defined below[34] is a good function:

$$\begin{aligned} S(x) &= C \exp\left(\frac{1}{1-x^2}\right) \quad -1 < x < 1 \\ &= 0 \quad |x| \geq 1 \end{aligned} \tag{6.4}$$

$$\text{with} \quad \int_{-\infty}^{\infty} dx S(x) = 1. \tag{6.5}$$

Solution: We note by inspection that the function is infinitely differentiable for all x except possibly at $|x| = 1$. It is further trivially bounded at infinity as it vanishes there. We thus have to check the nth order differentiability at $|x| = 1$. We note that $S(x)$ vanishes faster than any power of $(x \mp 1)$ as $x \to \pm 1^{\mp}$. Moreover, the derivatives of $S(x)$ necessarily contain a factor of $S(x)$. It is easy to verify that as $x \to \pm 1^{\mp}$, each derivative of $S(x)$ vanishes and hence is continuous across $x = \pm 1$.

6.2.2 Properties of good and fairly good functions

There are several properties of good and fairly good functions which follow from their respective definitions, and in some cases the use of the Leibniz rule. In the following, let $g(x)$, $g_1(x)$ and $g_2(x)$ denote good functions and $f(x)$, $f_1(x)$ and $f_2(x)$ denote fairly good functions.

Then, we can easily establish the following properties.

(a) Each derivative $g^{(n)}(x)$ is a good function.
(b) Each derivative $f^{(n)}(x)$ is a fairly good function.
(c) $f(x)g(x)$ is a good function.
(d) $(g_1 + g_2)$ and $(g_1 g_2)$ are good functions.
(e) $(f_1 + f_2)$ and $(f_1 f_2)$ are fairly good functions.
(f) The integrals $\int_{-\infty}^{\infty} g(x)dx$ and $\int_{-\infty}^{\infty} |g(x)|dx$ exist.

[34] We shall call it the 'smudge' function and use it in Section 6.3.2.

(g) The Fourier transform of $g(x)$ defined by

$$\tilde{g}(k) = \frac{1}{\sqrt{2\pi}} \int_{-\infty}^{\infty} dx g(x) \exp(-ikx) \tag{6.6}$$

exists. To prove this we note that

$$\left| \int_{-\infty}^{\infty} dx g(x) \exp(-ikx) \right| \leq \int_{-\infty}^{\infty} dx |g(x)| |\exp(-ikx)|$$
$$\leq \int_{-\infty}^{\infty} |g(x)| dx < \infty$$

from (f).

(h) The integral in the definition for the Fourier transform of a good function of equation (6.6) can be differentiated within with respect to k an arbitrary number of times. For proof, the reader is referred to results in the Appendix.

(i) Let $f(x)$ and $g(k)$ be two good functions of x and k respectively; and let $h(x, k)$ be a bounded function for $-\infty < x, k < \infty$. Then the order of integration in

$$\int_{-\infty}^{\infty} dx f(x) \int_{-\infty}^{\infty} dk g(k) h(x, k)$$

can be interchanged:

$$\int_{-\infty}^{\infty} dx f(x) \int_{-\infty}^{\infty} dk g(k) h(x, k) = \int_{-\infty}^{\infty} dk g(k) \int_{-\infty}^{\infty} dx f(x) h(x, k)$$

For proof, please refer to the results in the Appendix.

The proofs of (a)–(e) are left as an exercise (see Exercise 6.1).

The proof of (f) proceeds as follows: The fact that $g(x)$ is a good function allows us to place a bound

$$|g(x)| < \frac{A}{|x|^2} \quad \text{for} \quad |x| > L$$

with some A and L, $(0 < A, L < \infty)$, where we have (arbitrarily) chosen $N = 2$ in (6.2).

We can now express

$$\int_{-\infty}^{\infty} g(x) dx = \int_{-\infty}^{-L} g(x) dx + \int_{-L}^{L} g(x) dx + \int_{L}^{\infty} g(x) dx.$$

As $g(x)$ is a continuous function over the finite interval $(-L, L)$, there exists M such that $|g(x)| < M < \infty$. Then,

$$\int_{-L}^{L} g(x)dx < \int_{-L}^{L} |g(x)|dx < 2ML.$$

Also,

$$\left|\int_{L}^{\infty} g(x)dx\right| < \int_{L}^{\infty} |g(x)|dx < \int_{L}^{\infty} \frac{A}{|x|^2}dx < \frac{A}{L}.$$

Similarly, we have

$$\left|\int_{-\infty}^{-L} g(x)dx\right| < \int_{-\infty}^{-L} |g(x)|dx < \frac{A}{L}.$$

Thus,

$$\left|\int_{-\infty}^{\infty} g(x)dx\right| \leq \int_{-\infty}^{\infty} |g(x)|dx < 2ML + \frac{2A}{L} < \infty.$$

Corollary Making use of the properties (c) and (d), we see that for f, g, g_1 and g_2 as defined earlier, the integrals (i) $\int_{-\infty}^{\infty} f(x)g(x)dx$ and (ii) $\int_{-\infty}^{\infty} g_1(x)g_2(x)dx$ exist.

Comments: Note that the infinite differentiability of the Fourier transform of $g(x)$ under the integral sign has resulted from the restriction placed on its asymptotic behaviour. This property regarding the differentiability of the Fourier transform is very important for the use of Fourier representations in (i) solving differential equations and (ii) for Green's functions. We would not be able to use these techniques rigorously unless they are interpreted in the language of generalised functions. Further comments are found in Section 6.5

6.2.3 Definition of a generalised function and the Dirac delta function

Consider a sequence of good functions $\{\alpha_1(x), \alpha_2(x), \ldots, \alpha_n(x), \ldots\}$ such that the limit of the Riemann integral

$$\int_{-\infty}^{\infty} \alpha_n(x)g(x)dx$$

exists as $n \to \infty$ for every good function $g(x)$. Then a generalised function $\chi(x)$ is defined as the sequence

$$\{\alpha_1(x), \alpha_2(x), \ldots, \alpha_n(x), \ldots\}.$$

Further, we define the integral of a generalised function $\chi(x)$ with a good function $g(x)$ as

$$\int_{-\infty}^{\infty} \chi(x)g(x)dx \equiv \lim_{n\to\infty} \int_{-\infty}^{\infty} \alpha_n(x)g(x)dx.$$

We note that the integral on the right-hand side is the usual Riemann integral where as the integral on the left-hand side is not. The relation above *defines* what we mean by this integral on the left-hand side.

We introduce the Dirac delta function through the example below.

Solved problem 6.3 Show that the sequence $\{f_n(x); n = 1, 2, \ldots\}$ with

$$f_n(x) = \frac{n}{\sqrt{\pi}} e^{-n^2x^2}$$

defines a generalised functions $\delta(x)$. Verify explicitly that for any good function $g(x)$,

$$\int_{-\infty}^{\infty} \delta(x)g(x)dx = g(0).$$

(The above generalised function defines what is called the Dirac δ-function.)

Solution: We note that $f_n(x)$ is infinitely differentiable. Moreover, it is easy to verify the requirement of the asymptotic behaviour on $f_n(x)$ and its derivatives of all orders. (This happens primarily because of the overall factor $e^{-n^2x^2}$ present in each of these).

Thus, $\{f_n(x)\}$ constitutes a sequence of good functions.

We have to further establish that

$$\lim_{n\to\infty} \int_{-\infty}^{\infty} f_n(x)g(x)dx \equiv \lim_{n\to\infty} C_n$$

exists for every good function $g(x)$. We note that $f_n(x)g(x)$ is a good function ($F \bigotimes G \subseteq G$) and hence C_n exists. (See property (f) in Section 6.2.2.) To show that the sequence $\{C_n\}$ has a limit, we consider

$$\begin{aligned}
C_n &= \frac{n}{\sqrt{\pi}} \int_{-\infty}^{\infty} e^{-n^2x^2} g(x)dx = \frac{1}{\sqrt{\pi}} \int_{-\infty}^{\infty} e^{-y^2} g\left(\frac{y}{n}\right) dy \\
&= \frac{1}{\sqrt{\pi}} \int_{-\infty}^{\infty} e^{-y^2} g(0)dy + \frac{1}{\sqrt{\pi}} \int_{-\infty}^{\infty} e^{-y^2} \left[g\left(\frac{y}{n}\right) - g(0)\right] dy \\
&= g(0) + \frac{1}{\sqrt{\pi}} \int_{-\infty}^{\infty} e^{-y^2} \left[g\left(\frac{y}{n}\right) - g(0)\right] dy.
\end{aligned}$$

Now $g'(x)$ is a good function. Hence there exist B and L such that

$$|g'(x)| < B \quad \text{for} \quad |x| > L.$$

Also, $g'(x)$ is a continuous function over $(-L, L)$. Hence there exists M' such that

$$|g'(x)| < M' \quad \text{for} \quad -L \leq x \leq L.$$

We take M to be the greater of M' and B. Then we can write

$$|g'(x)| < M \qquad -\infty < x < \infty.$$

We can then write

$$\left| g\left(\frac{y}{n}\right) - g(0) \right| < M \left| \frac{y}{n} \right|.$$

Then,

$$|C_n - g(0)| < \frac{1}{\sqrt{\pi}} \int_{-\infty}^{\infty} e^{-y^2} |g\left(\frac{y}{n}\right) - g(0)| dy$$

$$< \frac{M}{\sqrt{\pi} n} \int_{-\infty}^{\infty} |y| e^{-y^2} dy = \frac{(\text{const})}{n} \to 0 \quad \text{as} \quad n \to \infty.$$

Thus, $\lim_{n\to\infty} C_n$ exists and is equal to $g(0)$ for any good function $g(x)$.

$$\int_{-\infty}^{\infty} \delta(x) g(x) dx \equiv \lim C_n = g(0).$$

Solved problem 6.4 Consider the sequence $\{t_1(x), t_2(x), \ldots, t_n(x), \ldots\}$ with $t_n(x) = e^{-x^2/n^2} F_n(x) \equiv e^{-x^2/n^2} \int_{-\infty}^{x} dy \frac{n}{\sqrt{\pi}} e^{-n^2 y^2}$. Show that this represents a generalised function (denoted by $\Theta(x)$).

It is easy to verify that t_n is infinitely differentiable. Moreover, the factor e^{-x^2/n^2} ensures that t_n and all its derivatives vanish sufficiently rapidly at infinity. Consider further,

$$\int_{-\infty}^{\infty} dx t_n(x) g(x)$$

$$= \int_{-\infty}^{\infty} dx g(x) e^{-x^2/n^2} \int_{-\infty}^{x} dy \frac{n}{\sqrt{\pi}} e^{-n^2 y^2}$$

$$= -\int_{-\infty}^{\infty} dx \frac{d}{dx} \left[\int_{x}^{\infty} dz g(z) e^{-x^2/n^2} \int_{-\infty}^{x} dy \frac{n}{\sqrt{\pi}} e^{-n^2 y^2} \right]$$

$$+ \int_{-\infty}^{\infty} dx \left[G(x) \frac{d}{dx} \left[e^{-x^2/n^2} \int_{-\infty}^{x} dy \frac{n}{\sqrt{\pi}} e^{-n^2 y^2} \right] \right].$$

In the second integral above, we have introduced the notation $G(x) \equiv \int_x^\infty dz g(z)$. The first integral on the right-hand side vanishes because the factor $e^{\frac{-x^2}{n^2}}$ vanishes at the boundaries while the remaining factors are bounded there. The second integral can be written as

$$\int_{-\infty}^{\infty} dx G(x) \frac{d}{dx}\left[e^{-x^2/n^2} \int_{-\infty}^{x} dy \frac{n}{\sqrt{\pi}} e^{-n^2 y^2}\right]$$

$$= \int_{-\infty}^{\infty} dx G(x) e^{-x^2/n^2} \frac{n}{\sqrt{\pi}} e^{-n^2 x^2} + \int_{-\infty}^{\infty} dx G(x) F_n(x) \frac{d}{dx}[e^{-n^2 x^2}]$$

$$= \int_{-\infty}^{\infty} dx G(x) \frac{n}{\sqrt{\pi}} e^{-n^2 x^2} + \int_{-\infty}^{\infty} dx G(x) [e^{-x^2/n^2} - 1] \frac{n}{\sqrt{\pi}} e^{-n^2 x^2}$$

$$+ \int_{-\infty}^{\infty} dx G(x) F_n(x) \frac{d}{dx}[e^{-n^2 x^2}].$$

We note that as $n \to \infty$, the first integral approaches $G(0)$. The remaining integrals can be easily shown to vanish as $n \to \infty$ using

(i) $[-e^{-x^2/n^2} + 1] < \frac{x^2}{n^2}$ for all x;
(ii) $G(x) < \infty$;
(iii) $\left|\frac{d}{dx}[e^{-x^2/n^2}]\right| < \sqrt{\frac{2}{e}}\frac{1}{n}$.

We leave the proof as an exercise.

Solved problem 6.5 Let $f(x)$ be any fairly good function. Then, the sequence $\{f(x)e^{-x^2/n^2}; n = 1, 2, \ldots\}$ defines a generalised function. We leave the proof as an exercise (see Exercise 6.11).

6.3 Relation between an ordinary function and a 'generalised function'

The applicability of the notion of a generalised function in physics may at first look too restricted. It employs the notion of good and fairly good functions which are very restricted classes of functions. It may appear that the functions that we deal with ordinarily in physics[35] are out of scope of this treatment, and hence this treatment would find few applications.

We shall however demonstrate that given a fairly general function $h(x)$ of x (we shall be precise about it soon), it can be replaced by a generalised function *in the sense to be defined*

[35]Recall that a square-well potential, a step potential, charge density of a uniformly charged sphere, etc., are commonly occurring discontinuous functions while a saw-tooth pulse has its first order derivative discontinuous; slope of wave functions and electromagnetic fields at boundaries can be discontinuous.

below. In particular, $h(x)$ can have (finitely many) finite discontinuities and quite 'liberal' asymptotic behaviour.

Conversely, we shall also note that the elaborate language of generalised functions may often not be required; we shall see this by making a correspondence of a generalised function to an ordinary function wherever it is possible.

6.3.1 Local value of a generalised function

Let $\Phi(x)$ be a generalised function and let $f(x)$ be a continuous function defined over the interval (a, b). Then, we say that the generalised function $\Phi(x)$ has a local value given by $f(x)$ in (a, b) if for *every* good function $g(x)$ vanishing outside the interval (a, b)[36] the following equality is fulfilled.

$$\int_{-\infty}^{\infty} dx g(x)\Phi(x) = \int_{a}^{b} dx g(x) f(x)$$

We note that the left-hand side is an integral of a generalised function, while the right-hand side is a Riemann integral over the interval $[a, b]$. We express this by saying that

$$\Phi(x) = f(x) \qquad a < x < b.$$

Examples

(a) Consider the Dirac delta function $\delta(x)$. For every good function $g(x)$ that vanishes outside the interval (ϵ, ∞), with $\epsilon > 0$, we have

$$\int_{-\infty}^{\infty} dx g(x)\delta(x) = 0 = \int_{\epsilon}^{\infty} dx g(x).0.$$

We can thus conclude that

$$\delta(x) = 0 \quad 0 < \epsilon < x < \infty.$$

A similar argument allows one to conclude (taking ϵ arbitrarily small)

$$\delta(x) = 0 \quad x \neq 0.$$

(b) Consider $\Theta(x)$ defined earlier in Section 6.2.3. It is easy to verify that (see Exercise 6.10) for every good function vanishing outside the interval $(0, \infty)$,

$$\int_{-\infty}^{\infty} dx g(x)\Theta(x) = \int_{0}^{\infty} dx g(x).1$$

[36] In such a case we say that this is a function with a *support* in the interval (a, b).

implying

$$\Theta(x) = 1 \qquad 0 < x < \infty.$$

A similar argument allows one to deduce

$$\Theta(x) = 0 \qquad -\infty < x < 0.$$

(c) Consider a fairly good function $f(x)$ and a generalised function $\Phi(x)$ defined by the sequence $\{f_n(x) \equiv f(x)e^{-\frac{x^2}{n^2}}\}$. Then, for every good function (see Exercise 6.3),

$$\int_{-\infty}^{\infty} dx g(x)\Phi(x) = \int_{-\infty}^{\infty} dx g(x) f(x).$$

Thus, we have

$$\Phi(x) = f(x) \qquad -\infty < x < \infty.$$

A criterion for $\Phi(x)$ to have a local value

Let $\{f_1(x), f_2(x), \ldots, f_n(x), \ldots\}$ be some sequence of good functions representing the generalised function $\Phi(x)$. Consider also a continuous function $f(x)$. Suppose there exits a neighbourhood $|x - x_0| < \delta > 0$ such that given an arbitrarily small $\epsilon > 0$,

$$|f_n(x) - f(x)| < \epsilon$$

whenever $n > N_0$ *for every* x in this neighbourhood, then $\Phi(x_0) = f(x_0)$.

Proof: Consider an arbitrary good function $g(x)$ with a support on the interval $[x_0-\delta, x_0+\delta]$. Then,

$$\begin{aligned}\int_{-\infty}^{\infty} dx \Phi(x) g(x) &= \lim_{n\to\infty} \int_{-\infty}^{\infty} dx f_n(x) g(x) \\ &= \int_{-\infty}^{\infty} dx f(x) g(x) \\ &\quad + \lim_{n\to\infty} \int_{-\infty}^{\infty} dx [f_n(x) - f(x)] g(x), \end{aligned} \tag{6.7}$$

and for an arbitrarily small ϵ, there exists an N_0 such that

$$\left| \int_{-\infty}^{\infty} dx [f_n(x) - f(x)] g(x) \right| < \epsilon \int_{-\infty}^{\infty} dx |g(x)| < \epsilon M$$

for all $n > N_0$ [here, $M < \infty$]. Thus, the second integral on the right-hand side of (6.7) approaches zero, proving the result.

6.3.2 Replacing an ordinary function by a generalised function

Let $h(x)$ be a function defined over the real axis

(a) having (finitely many) finite discontinuities at $x = x_{i0}$, $i = 1, 2, 3, \ldots, N$;
(b) with $\int_{-\infty}^{\infty} dx \left|\frac{h(x)}{(1+x^2)^N}\right| < \infty$ for some nonnegative N.

We will to show that we can replace this function by a generalised function Φ in the sense that for every good function $g(x)$,

$$\int_{-\infty}^{\infty} dx g(x)\Phi(x) = \int_{-\infty}^{\infty} dx g(x)h(x)$$

exists. We shall further show that $\Phi(x)$ has the local value

$$\Phi(x) = h(x) \quad x \neq x_{i0} \tag{6.8}$$

and the sequence defining $\Phi(x)$ converges to $1/2[h(x_{i0}^+) + h(x_{i0}^-)]$ at the discontinuities. The latter result will have relevance when we discuss the Fourier transforms.

This will be shown by explicit construction of $\Phi(x)$. We consider the function $\eta_n(x)$ defined by

$$\eta_n(x) \equiv \int_{-\infty}^{\infty} dt h(t) n e^{-\frac{t^2}{n^2}} S[n(t-x)],$$

where $S[n(t-x)]$ is the smudge function defined earlier in (6.4)–(6.5) that it is nonzero only over a small interval $(x - 1/n, x + 1/n)$. Before we proceed, we note that, qualitatively speaking, the right-hand side above takes a weighted average of $h(t)$ over a small interval around $t = x$ with (an infinitely) smooth function $S[n(t-x)]$. The interval itself shrinks to zero as $n \to \infty$. The averaging process smoothens any discontinuous behaviour that the function $h(t)$ may have. Moreover, the presence of the factor e^{-t^2/n^2} makes the function fall rapidly of as $|x| \to \pm\infty$ as required of a good function. We note that

$$\begin{aligned}
|\eta_n^{(p)}(x)| &= \left|\int_{-\infty}^{\infty} dt h(t)(-n)^p S^{(p)}[n(t-x)] n e^{-t^2/n^2}\right| \\
&= \left|\int_{-\infty}^{\infty} dt h(t)(-n)^p S^{(p)}[n(t-x)] n e^{-t^2/n^2}\right| \\
&< n^{p+1} M e^{-(|x|-1)^2/n^2}[1 + (|x|+1)^2]^N \\
&\quad \times \int_{-\infty}^{\infty} dt |h(t)|(1+t^2)^{-N}.
\end{aligned}$$

In arriving at the above, we have made use of the following.

(a) Since $S^p[n(t-x)]$ is a good function, there exists an M such that $|S^p[n(t-x)]| < M$ for all t.
(b) We have also used $e^{-t^2/n^2} < e^{-(|x|-1)^2/n^2}$ within the range of integration.
(c) We have used $(1+t^2)^N < [1+(|x|+1)^2]^N$ for a nonnegative N for all n within the range of integration.
(d) N is an arbitrary nonnegative number for which the integral

$$\int_{-\infty}^{\infty} dt|h(t)|(1+t^2)^{-N}$$

exists.

We thus note from the above bound that because of the presence of the exponential, $\eta_n^{(p)}(x)$ vanishes faster than any power of $1/x$ provided $\int_{-\infty}^{\infty} dt|h(t)|(1+t^2)^{-N}$ exists for some N. Thus, $\eta_n(x)$ are good functions. Now,

$$\begin{aligned}
\eta_n(x) &= \int_{-\infty}^{\infty} ndth\,(t)\,e^{-t^2/n^2} S[n(t-x)] \\
&= \int_{-1}^{1} d\xi h\left(x+\frac{\xi}{n}\right) e^{-\left(x+\frac{\xi}{n}\right)^2/n^2} S\,(\xi) \\
&= \int_{0}^{1} d\xi h\left(x+\frac{\xi}{n}\right) e^{-\left(x+\frac{\xi}{n}\right)^2/n^2} S\,(\xi) \\
&\quad + \int_{-1}^{0} d\xi h\left(x+\frac{\xi}{n}\right) e^{-\left(x+\frac{\xi}{n}\right)^2/n^2} S\,(\xi)\,.
\end{aligned} \tag{6.9}$$

We shall now assume that n is large enough so that there are no discontinuities of $h(x)$ in the open intervals $(x, x+\frac{\xi}{n})$ and $(x-\frac{\xi}{n}, x)$. Now,

$$\begin{aligned}
&\int_{0}^{1} d\xi h\left(x+\frac{\xi}{n}\right) e^{-\left(x+\frac{\xi}{n}\right)^2/n^2} S\,(\xi) \\
&= \int_{0}^{1} d\xi h(x^+) e^{-x^2/n^2} S\,(\xi) + \int_{0}^{1} d\xi h(x^+) \left[e^{-\left(x+\frac{\xi}{n}\right)^2/n^2} - e^{-x^2/n^2}\right] S(\xi) \\
&\quad + \int_{0}^{1} d\xi \left[h\left(x+\frac{\xi}{n}\right) - h(x^+)\right] e^{-\left(x+\frac{\xi}{n}\right)^2/n^2} S\,(\xi) \\
&= \frac{1}{2} h(x^+) e^{-x^2/n^2} + h(x^+) \int_{0}^{1} d\xi \left[e^{-\left(x+\frac{\xi}{n}\right)^2/n^2} - e^{-x^2/n^2}\right] S(\xi) \\
&\quad + \int_{0}^{1} d\xi \left[h\left(x+\frac{\xi}{n}\right) - h(x^+)\right] e^{-\left(x+\frac{\xi}{n}\right)^2/n^2} S\,(\xi)\,.
\end{aligned} \tag{6.10}$$

Thus,

$$\left|\int_0^1 d\xi h\left(x+\frac{\xi}{n}\right)e^{-\left(x+\frac{\xi}{n}\right)^2/n^2}S(\xi)-\frac{1}{2}h(x^+)e^{-x^2/n^2}\right|$$
$$\le \int_0^1 d\xi\left|\left[h\left(x+\frac{\xi}{n}\right)-h(x^+)\right]\right|e^{-\left(x+\frac{\xi}{n}\right)^2/n^2}S(\xi)$$
$$+h(x^+)\left[e^{-\left(|x|-\frac{1}{n}\right)^2/n^2}-e^{-x^2/n^2}\right]\int_0^1 d\xi S(\xi)$$
$$<\int_0^1 d\xi\left|\left[h\left(x+\frac{\xi}{n}\right)-h(x^+)\right]\right|S(\xi)$$
$$+\frac{1}{2}h(x^+)\left[e^{-\left(|x|-\frac{1}{n}\right)^2/n^2}-e^{-x^2/n^2}\right]. \tag{6.11}$$

Now, since $h(x)$ is continuous in $(x, x+\frac{\xi}{n})$,

$$\left|h\left(x+\frac{\xi}{n}\right)-h(x^+)\right|<M\left|\frac{\xi}{n}\right|<\frac{M}{n}$$

for some M. Thus,

$$\left|\int_0^1 d\xi h\left(x+\frac{\xi}{n}\right)e^{-\left(x+\frac{\xi}{n}\right)^2/n^2}S(\xi)-\frac{1}{2}h(x^+)e^{-x^2/n^2}\right|$$
$$<\frac{M}{2n}+1/2h(x^+)\left[e^{-\left(|x|-\frac{1}{n}\right)^2/n^2}-e^{-x^2/n^2}\right]\to 0 \tag{6.12}$$

as $n\to\infty$. We can treat the integral in (6.9) likewise. Then, we find that

$$\lim_{n\to\infty}\eta_n(x)=\frac{1}{2}[h(x^+)+h(x^-)].$$

In a similar manner we can show that the limit

$$\lim_{n\to\infty}\int_{-\infty}^{\infty}dx\eta_n(x)g(x)$$

exists for every good function $g(x)$ and equals

$$\int_{-\infty}^{\infty}dxh(x)g(x).$$

The above proof is left as an exercise (use the same techniques used above in proving $\eta_n(x) \to 1/2[h(x^-) + h(x^+)$. See Exercise 6.9). This proves that the sequence $\{\eta_1(x), \eta_2(x), \ldots, \eta_n(x), \ldots\}$ defines a generalised function corresponding to the ordinary function $h(x)$.

6.4 Operations on the Generalised Functions

In this section we shall introduce several operations on generalised functions. The discussion is simple and most proofs are obvious (see Exercise 6.4).

(a) Addition of generalised functions Let Φ and Ψ be two generalised functions represented respectively by the sequences of good functions

$$\Phi = \{f_1, f_2, \ldots, f_n, \ldots\}, \tag{6.13}$$

$$\Psi = \{g_1, g_2, \ldots, g_n, \ldots\}. \tag{6.14}$$

Then the sequence $\{f_1 + g_1, f_2 + g_2, \ldots, f_n + g_n, \ldots\}$ represents a generalised function. We denote it by $\Phi + \Psi$. This is referred to as the addition of generalised functions.

(b) Multiplication of a generalised function by a constant Let Φ be a generalised function represented as in (a). Then, the sequence

$$\{\alpha f_1, \alpha f_2, \ldots, \alpha f_n, \ldots\}$$

represents a generalised function denoted by $\alpha\Phi$.

Further, let Φ and Ψ be two generalised functions and let α and β be complex constants. Then the sequence, $\{\alpha f_1 + \beta g_1, \alpha f_2 + \beta g_2, \ldots, \alpha f_n + \beta g_n, \ldots\}$ represents a generalised function $[\alpha\Phi + \beta\Psi]$.

(c) Multiplication of a generalised function by a fairly good function Let Ψ be a generalised function represented by a sequence

$$\{g_1, g_2, \ldots, g_n, \ldots\}$$

and let $F(x)$ be a fairly good function. Then the sequence

$$\{Fg_1, Fg_2, \ldots, Fg_n, \ldots\}$$

represents a generalised function (which we will denote by $F\Psi$).

Proof:

(a) We note that the sequence under consideration is a sequence of good functions.

(b) For any good function $G(x)$, consider the limit

$$\lim_{n\to\infty}\int_{-\infty}^{\infty} dx F(x) g_n(x) G(x) = \lim_{n\to\infty}\int_{-\infty}^{\infty} dx g_n(x)\{F(x)G(x)\}.$$

The last limit exists since $\{F(x)G(x)\}$ is a good function and $\{g_1, g_2, \dots, g_n, \dots\}$ is a sequence representing a generalised function Ψ. Hence, the sequence $\{Fg_1, Fg_2, \dots, Fg_n, \dots\}$ represents a generalised function.

(d) Differentiation of a generalised function Let a generalised function Φ be represented by a sequence of good functions $\{f_1, f_2, \dots, f_n, \dots\}$.

Then

$$\left\{\frac{df_1}{dx}, \frac{df_2}{dx}, \dots, \frac{df_n}{dx}, \dots\right\}$$

is a sequence of good functions that represents a generalised function. We denote it by $\frac{d\Phi}{dx}$.

Proof: We already know that the derivative of a good function is a good function. Thus the sequence

$$\left\{\frac{df_1}{dx}, \frac{df_2}{dx}, \dots, \frac{df_n}{dx}, \dots\right\}$$

is a sequence of good functions. Now let $g(x)$ be a good function. Consider

$$\lim_{n\to\infty}\int_{-\infty}^{\infty} dx \frac{df_n}{dx} g(x) = -\lim_{n\to\infty}\int_{-\infty}^{\infty} dx \frac{dg}{dx} f_n(x). \tag{6.15}$$

Since $\frac{dg}{dx}$ is a good function, the last limit exists for an arbitrary good function $g(x)$. Hence the sequence

$$\left\{\frac{df_1}{dx}, \frac{df_2}{dx}, \dots, \frac{df_n}{dx}, \dots\right\}$$

represents a generalised function.

Comments:

1. It follows that a generalised function can be differentiated an arbitrary number of times.
2. The differentiation of a generalised function follows the rules (see Exercise 6.2)

$$\frac{d}{dx}[\alpha\Phi + \beta\Psi] = \left[\alpha\frac{d\Phi}{dx} + \beta\frac{d\Psi}{dx}\right],$$

$$\frac{d}{dx}[f(x)\Phi(x)] = \frac{df}{dx}\Phi(x) + f(x)\frac{d\Phi}{dx}.$$

(e) Equality of two generalised functions Consider two generalised functions $\Phi(x)$ and $\Psi(x)$ such that

$$\int_{-\infty}^{\infty} dx\Phi(x)g(x) = \int_{-\infty}^{\infty} dx\Psi(x)g(x)$$

holds for every good function $g(x)$. Then we say that the two generalised functions are equal.

In particular, $\{f_n(x); n = 1, 2, \dots\}$ and $\{h_n(x); n = 1, 2, \dots\}$ be two sequences of good functions such that for every good function $g(x)$, the following limits exist and are equal.

$$\lim_{n\to\infty} \int_{-\infty}^{\infty} dxg(x)f_n(x) = \lim_{n\to\infty} \int_{-\infty}^{\infty} dxg(x)h_n(x)$$

Then the two sequences represent the same generalised function.

(f) We note that, in general, the product of generalised functions cannot be given a definition.

6.5 Fourier Transforms of Generalised Functions

In this section we shall deal with the Fourier transform of generalised functions. In order to do so, it is first necessary to deal with the Fourier transform of good functions as the former is a sequence of these. The Fourier transform of a good function $g(x)$ is defined by

$$\tilde{g}(k) = \frac{1}{\sqrt{2\pi}} \int_{-\infty}^{\infty} dxg(x)\exp(-ikx), \tag{6.16}$$

which we have already shown to exist. We shall first show that $\tilde{g}(k)$ is itself a good function. To show this, we need to show that $\tilde{g}(k)$ and its arbitrary order derivatives $\frac{d^p\tilde{g}(k)}{dk^p}$ exist and vanish faster than any power of $\frac{1}{|k|}$ as $|k| \to \infty$. As shown in Section 6.2.2, the integral on the right-hand side of (6.16) can be differentiated inside the integral (see Exercise 6.5). Thus,

$$\frac{d^p\tilde{g}(k)}{dk^p} = \frac{1}{\sqrt{2\pi}} \int_{-\infty}^{\infty} dxg(x)(-ix)^p \exp(-ikx). \tag{6.17}$$

To show that this falls off faster than any power, we integrate it by parts N times, where N is arbitrary positive integer. Each time the 'surface' terms are proportional to[37] the value of $x^p g(x)$ or one of its derivatives $\frac{d^q[x^p g(x)]}{dx^q}$ at $\pm\infty$. But all of these are good functions (see the

[37] This, in fact, explains the necessity of defining a good function in the restrictive manner that we did. Without the property regarding the asymptotic behaviour, the differentiation under integral sign an arbitrary number of times would not have been possible.

properties of good functions in Section 6.2.2) and therefore the surface terms vanish. Thus, we have

$$\frac{d^p \tilde{g}(k)}{dk^p} = (-i)^p(-i)^N \frac{1}{k^N}\frac{1}{\sqrt{2\pi}}\int_{-\infty}^{\infty} dx \frac{d^N[x^p g(x)]}{dx^N} \exp(-ikx). \tag{6.18}$$

Now, the integral in the right-hand side above exists, since $\frac{d^N[x^p g(x)]}{dx^N}$ is a good function itself and is moreover bounded by a k-independent constant

$$\left|\int_{-\infty}^{\infty} dx \frac{d^N[x^p g(x)]}{dx^N} \exp(-ikx)\right| \leq \int_{-\infty}^{\infty} dx \left|\frac{d^N[x^p g(x)]}{dx^N}\right| < \infty.$$

Thus, as $|k| \to \infty$, each of the derivatives $\frac{d^p \tilde{g}(k)}{dk^p}$ vanishes at least as fast as $|k|^{-N}$; this is valid for each p for any arbitrary positive N. Thus, $\tilde{g}(k)$ is a good function.

Example Consider $f_n(x) = \frac{n}{\sqrt{\pi}} \exp(-n^2x^2)$. Its Fourier transform $\tilde{f}_n(k)$ given by

$$\tilde{f}_n(k) = \frac{1}{\sqrt{2\pi}}\frac{n}{\sqrt{\pi}}\int_{-\infty}^{\infty} dx \exp(-ikx)\exp(-n^2x^2)$$

can be evaluated by completing the square in the exponent. This is found to be

$$\tilde{f}_n(k) = \frac{1}{\sqrt{2\pi}} \exp\left(-\frac{k^2}{4n^2}\right).$$

This is in turn a good function. In fact we note that the sequence of good functions

$$\{\tilde{f}_1(k), \tilde{f}_2(k), \ldots, \tilde{f}_n(k), \ldots\}$$

defines a generalised function. To see this, let $G(k)$ be any arbitrary good function. We consider

$$\lim_{n\to\infty} \int_{-\infty}^{\infty} dk G(k) \tilde{f}_n(k).$$

We can take the limit inside the integral sign (see Exercise 6.3) and employ $\lim_{n\to\infty} \tilde{f}_n(k) = \frac{1}{\sqrt{2\pi}}$. This implies that the above limit exists and equals $\frac{1}{\sqrt{2\pi}}\int_{-\infty}^{\infty} dk G(k)$. This proves the assertion[38].

---

[38]In this section, we shall often interchange the order of limits and/or integrations in various places. Please refer to Appendix A for a justification.

Theorem Let $f(x)$ be a good function with the Fourier transform

$$\tilde{f}(k) = \frac{1}{\sqrt{2\pi}} \int_{-\infty}^{\infty} dx f(x) \exp(-ikx).$$

Then, $f(x)$, the *inverse* Fourier transform of $\tilde{f}(k)$, is given by

$$f(x) = \frac{1}{\sqrt{2\pi}} \int_{-\infty}^{\infty} dk \tilde{f}(k) \exp(ikx).$$

Proof: Consider

$$\begin{aligned} f(x) &= \int_{-\infty}^{\infty} dx' f(x') \delta(x - x') \\ &= \lim_{n\to\infty} \int_{-\infty}^{\infty} dx' f(x') \frac{n}{\sqrt{\pi}} e^{-(x-x')^2 n^2} \\ &= \lim_{n\to\infty} \int_{-\infty}^{\infty} dx' f(x') \int_{-\infty}^{\infty} \frac{dk}{2\pi} e^{i(x-x')k} e^{\frac{-k^2}{4n^2}} \\ &= \lim_{n\to\infty} \int_{-\infty}^{\infty} \frac{dk}{\sqrt{2\pi}} e^{ikx} e^{\frac{-k^2}{4n^2}} \int_{-\infty}^{\infty} \frac{dx'}{\sqrt{2\pi}} f(x') e^{-ikx'} \\ &= \lim_{n\to\infty} \int_{-\infty}^{\infty} \frac{dk}{\sqrt{2\pi}} e^{ikx} e^{\frac{-k^2}{4n^2}} \tilde{f}(k) \\ &= \frac{1}{\sqrt{2\pi}} \int_{-\infty}^{\infty} dk e^{ikx} \tilde{f}(k), \end{aligned} \tag{6.19}$$

thus proving the result.

Parseval Identity

Let $f(x)$ and $F(x)$ be good functions with Fourier transforms $\tilde{f}(k)$ and $\tilde{F}(k)$. Then,

$$\int_{-\infty}^{\infty} dx f(x) F(x) = \int_{-\infty}^{\infty} dk \tilde{f}(k) \tilde{F}(-k). \tag{6.20}$$

Proof:

$$\begin{aligned} \int_{-\infty}^{\infty} dx f(x) F(x) &= \int_{-\infty}^{\infty} dx \frac{1}{\sqrt{2\pi}} \int_{-\infty}^{\infty} dk \tilde{f}(k) \exp(ikx) F(x) \\ &= \int_{-\infty}^{\infty} dk \tilde{f}(k) \frac{1}{\sqrt{2\pi}} \int_{-\infty}^{\infty} dx \exp(ikx) F(x) \\ &= \int_{-\infty}^{\infty} dk \tilde{f}(k) \tilde{F}(-k), \end{aligned} \tag{6.21}$$

which proves the result.

Fourier transform of a generalised function

Let Φ be a generalised function defined by sequence of good functions $\{f_1(x), f_2(x), \ldots, f_n(x), \ldots\}$. We now consider the sequence of the Fourier transforms of these, viz., $\{\tilde{f}_1(k), \tilde{f}_2(k), \ldots, \tilde{f}_n(k), \ldots\}$. As shown, this itself constitutes a sequence of good functions. We would like to first show that this sequence defines a generalised function $\tilde{\Phi}(k)$. To show this, we consider first the sequence of integrals

$$I_n \equiv \int_{-\infty}^{\infty} dk \tilde{G}(k) \tilde{f}_n(k)$$

with $\tilde{G}(k)$ an arbitrary good function and show that it approaches a limit as $n \to \infty$. To see this, we use the Parseval identity and find that

$$I_n = \int_{-\infty}^{\infty} dx G(-x) f_n(x),$$

where $G(-x)$ is the inverse Fourier transform of $G(k)$ and thus a good function. Now, since $\{f_n(x)\}$ is a sequence of good functions corresponding to a generalised function, limit of the above integral is known to exist, thus proving the result.

We close by making a remark about the Fourier transform of an ordinary function. Let $h(x)$ be an ordinary function with a finitely many finite discontinuities at $x = x_{i0}, i = 1, 2, \ldots$ and which has a Fourier transform. Let $\Phi(x)$ denote the generalised function associated with $h(x)$. If the Fourier transform of $\Phi(x)$ is $\tilde{\Phi}(k)$, then the *inverse* Fourier transform of $\tilde{\Phi}(k)$ will be $\Phi(x)$ again. This will have the local value of $h(x)$ at any point x where $h(x)$ is continuous, and the sequence defining $\Phi(x)$ will converge to $1/2[h(x_{i0}^+) + h(x_{i0}^-)]$ at the discontinuity at x_{i0} as shown in Section 6.3.2.

Exercises

6.1 Prove the properties of the good and the fairly good functions listed in Section 6.2.2.

6.2 Show that the differentiation of a generalised function follows the rules

$$\frac{d}{dx}[\alpha\Phi + \beta\Psi] = \left[\alpha\frac{d\Phi}{dx} + \beta\frac{d\Psi}{dx}\right]$$

and

$$\frac{d}{dx}[f(x)\Phi(x)] = \frac{df}{dx}\Phi(x) + f(x)\frac{d\Phi}{dx}.$$

6.3 Prove that, for any good function $G(k)$, the limit $n \to \infty$ can be taken inside the integral

$$\lim_{n\to\infty} \int_{-\infty}^{\infty} dk G(k) e^{\frac{-k^2}{4n^2}} = \int_{-\infty}^{\infty} dk G(k).$$

[Hint: Use $1 - e^{-t^2} < t^2$.]

6.4 Prove the properties of generalised functions stated in Section 6.4.

6.5 Show that for a good function $g(x)$, the integral

$$\hat{g}(\alpha) \equiv \int_{-\infty}^{\infty} dx g(x) f(\alpha, x)$$

can be differentiated with respect to α inside the integral any number of times if $f(\alpha, x)$ is a differentiable function of α for all x bounded as $x \to \infty$.

6.6 Verify that the function $S(x)$ of (6.5) is infinitely differentiable at $x = \pm 1$.

6.7 Suppose that a Green's function G(x,x') satisfying

$$\frac{d^2}{dx^2} G(x, x') + m^2 G(x, x') = \delta(x - x')$$

is to be solved by expressing it as the Fourier integral:

$$G(x, x') = \int_{-\infty}^{\infty} dk \tilde{G}(k) e^{ik(x-x')}.$$

Comment on the properties expected of $\tilde{G}(k)$ regarding integrability.

6.8 The Dirac delta-function $\delta(x)$ is often expressed as the limit $\epsilon \to 0^+$ of the following function:

$$\begin{aligned} f(x) &= \frac{1}{2\epsilon} \qquad -\epsilon < x < \epsilon \\ &= 0 \qquad \text{otherwise.} \end{aligned} \tag{6.22}$$

(a) Is $f(x)$ a good function? A fairly good function?

(b) Justify the conventional use of this definition of δ-function by constructing a sequence of *good* functions that resembles the above set of functions.

(c) Repeat this exercise for $f(x) = \frac{1}{\pi} \frac{\sin kx}{k}$ which is often used as a representation for the Dirac δ-function as $k \to \infty$.

6.9 Prove that for $\eta_n(x)$ defined in Section 6.3.2,

$$\lim_{n\to\infty} \int_{-\infty}^{\infty} dx \eta_n(x) g(x) = \int_{-\infty}^{\infty} dx h(x) g(x)$$

for every good function $g(x)$.

6.10 Consider $\Theta(x)$ defined earlier in Section 6.2.3. Show that for every good function $g(x)$ vanishing outside the interval $(0, \infty)$,

$$\int_{-\infty}^{\infty} dx g(x) \Theta(x) = \int_{0}^{\infty} dx g(x).1$$

6.11 Let $f(x)$ be a fairly good function. Then, show that the sequence $\{F_n(x) \equiv f(x)e^{-x^2/n^2}; n = 1, 2, \ldots\}$ defines a generalised function.

6.12 Show that the generalised function $\Theta(x)$ defined in Section 6.2.3 satisfies

$$\frac{d}{dx}\Theta(x) = \delta(x).$$

6.13 Let the Fourier transform of the generalised function $\Phi(x)$ be $\tilde{\Phi}(k)$. Then show that the Fourier transform of $\frac{d}{dx}\Phi(x)$ is $ik\tilde{\Phi}(k)$.

References

[1] Lighthill, M. J., *Introduction to Fourier Analysis and Generalised Functions*, Cambridge University Press, New York, 1958.

[2] Titchmarsh, E. C., *Introduction to the Theory of Fourier Integrals*, Chelsea Publishing Company, New York, 1986.

7 Group Theory

7.1 Introduction

The subject of group theory has found wide applications in all branches of physics. We shall elaborate the reason for this a little later. We shall first name a few of the important applications.

1. The theory of angular momentum upon which, in particular, the spectroscopy of atoms and nuclei is based
2. The symmetry groups of crystals and applications to the spectroscopy of crystalline solids
3. The spectroscopy of elementary particles (multiplet structure) – also based on symmetry groups
4. The fundamental theory of strong, week and electromagnetic interactions (the standard model) and its possible extensions has at its basis, a symmetry group called the 'gauge group'
5. The representations of the Lorentz group which determine the states elementary particles have

In this chapter we wish to limit ourselves to only an exposure to subject.

It is important to first understand why group theory has come to play an important role in physics. To begin with, we recall that a given physical system may have a number of symmetries. To name a few of the commonly occurring symmetries, we note that the dynamics of a large class of physical problems have symmetries under space and time translations. Dynamics of problems with central forces (such as an hydrogen atom) has invariance under rotations in the three-dimensional space. The Hamiltonian for an electron in presence of a crystalline solid has symmetries associated with that of the crystalline structure, characterised by what is called a 'point group'. Relativistic formulations (for example, the Dirac equation, classical electrodynamics) have a symmetry under the transformations of the Lorentz group that contains, in particular, the familiar Lorentz transformations. The use of group theory in physics is inextricably linked with the presence of symmetries (or approximate symmetries).

We shall now attempt to understand how the set of symmetries of a system conform to a mathematical structure we call a 'group' (to be introduced in the next section). Consider a quantum mechanical system with a Hamiltonian H. Let R_1 represent the operator corresponding to a symmetry operation for the system (a symmetry operation is an operation on the system that leaves the system unchanged). Then H is invariant under R_1: $R_1 H R_1^{-1} = H$ i.e., $R_1 H = H R_1$. Let R_2 be another symmetry operation for this physical system. Then, $R_2 H = H R_2$. Now, $R_1 R_2 H = R_1 H R_2 = H R_1 R_2$. Thus, the successive applications of two symmetry operations represented by $(R_1 R_2) \equiv R_3$ is also a symmetry operation for the system. If we interpret the symmetry operations from an *active* viewpoint (a symmetry operation affects the system rather than the reference frame), we know by construction how successive symmetry operations act on a physical system, i.e., we know their 'composition rule'. We know from experience that such an application of symmetry operations in succession is associative. Further, the 'identity operation' that leaves the physical system unchanged is a trivial symmetry operation. Thus, the set S of ALL the symmetry operations of a physical system will have the following properties:

(i) The existence of an associative composition rule (denoted henceforth by $\circ$)
(ii) $R_1, R_2 \in S \Rightarrow R_1 \circ R_2 \in S$
(iii) existence of an identity operation in S
(iv) $R \in S \Rightarrow R^{-1} \in S$

These are, in fact, the properties of the set of all symmetries of a system, and these are abstracted (from a physicist's view-point) into the basic definition of a *group* to be presented in the next section.

We remark that once we have abstracted a group of symmetries from some (physical) system, it exists as a mathematical entity irrespective of its connection with any physical system and can be employed in systems which may not have it as an exact symmetry group.

Now, having elucidated the connection between symmetries of a system and the group structure, we would like to know how the considerations of the group structure of symmetries become valuable in physics. In physics, barring some simple examples, most systems are complex and not amenable to direct methods or exact solutions. To illustrate this point, we shall take several examples from different branches of physics. Consider for example: (i) electronic levels in presence of crystalline field of a cubic crystal; (ii) energy levels of a nucleon in a nuclear potential well; (iii) masses and the other properties of the spinless mesons in elementary particle physics.

Each of the above is a complex physics problem whose exact analytic treatment may not be known or even possible. Under such circumstances, the group theoretical considerations become very useful and lead to several 'structural' results without even any detailed knowledge of the dynamics or the techniques for solution. For example, in case (i) we may not know the exact energy levels and their wave functions, but we may know the groups of energy levels which must be the degenerate and the relative intensities of groups of the spectral lines. In case (ii), assumption of a spherically symmetric potential enables one to know the degeneracies of states. One can then add to the potential terms that break this symmetry and know how this degeneracy is lifted. In case (iii), an assumption of an isospin symmetry allows one to conclude which of the mesons must have identical masses if the symmetry is exact. A further assumption of an approximate $SU(3)$ symmetry enables one to predict, in addition, relations between unequal masses.

Equally importantly, *deviations from a symmetry* (which are often found) can be conveniently formulated in the group theory framework.

We close by writing down several familiar physical systems and their symmetries.

1. Various molecules such as water (H_2O), ammonia (NH_3), benzene (C_6H_6) have symmetries under rotations, and reflections about appropriate planes and axes.
2. A cubic crystal has symmetry under certain rotations and reflections that keep a lattice site fixed. Various other crystal structures have similar symmetries.
3. The hydrogen atom Hamiltonian is invariant under arbitrary rotations about its centre.

7.2 Definitions and Properties

7.2.1 Definition of a group

An abstract group G is a set of elements $\{a, b, c, \ldots\}$ on which a 'law of multiplication' is defined (denoted by $\circ$) and which satisfies the following conditions[39]:

1. $a \in G, b \in G \Rightarrow a \circ b \in G$
2. The multiplication law is associative: $a \circ (b \circ c) = (a \circ b) \circ c$
3. There exists an element $e \in G$ such that $ae = a$ for every $a \in G$
4. For each $a \in G$, there exists an element $a^{-1} \in G$ such that $a \circ a^{-1} = e$

We hasten to add a number of elaborative comments.

Comments:

(a) There is no restriction on the number of elements of the set G which could either be finite or infinite. Accordingly, it is then known as a 'finite' group or an 'infinite' group.

(b) The condition (c) above defines e as a 'right identity'. We shall show shortly that it is also a 'left identity' and, in fact, the unique identity element of G. In a similar manner, we have defined a^{-1} as a 'right inverse' which as we shall see is the unique inverse of $a \in G$.

(c) For a given group, the multiplication law, also called 'the group multiplication' need not necessarily be commutative (i.e., the element $a \circ b$ need not generally be equal to $b \circ a$.)

(d) One should remember that group multiplication is a general and an abstract notion, that is, specific examples of it may not resemble any known form of (or conform to a conventional notion of) what we normally understand as 'multiplication'.

---

[39]In this chapter, we shall extensively use set-theoretic notations.

Henceforth, we shall generally omit the multiplication symbol $\circ$. Thus ab will stand for $a \circ b$.

We shall now present several examples which are specific realizations of the abstract notion of a group.

Examples of groups

1. Let $G \equiv$ set of all integers; and $\circ \equiv$ law of addition of numbers.
   Let the elements of G be (alternately) denoted by a_n, with a_n identified with the integer n, i.e., $(a_n \leftrightarrow n)$. Hence,

   $$G \equiv \{\ldots, a_{-2}, a_{-1}, a_0, a_1, a_2, \ldots\}.$$

   (a) $a_n \circ a_m \leftrightarrow n + m \leftrightarrow a_{n+m}$.
   (b) Associativity of $\circ$ is evident from the associativity of addition of numbers.
   (c) The identity element is $e \leftrightarrow a_0$: $a_n \circ a_0 \leftrightarrow n + 0 = n \leftrightarrow a_n$.
   (d) The inverse of a_n is a_{-n} since

   $$a_n \circ a_{-n} \leftrightarrow n + (-n) = 0 \leftrightarrow e.$$

   We express the group nature of this set G under the above law of group multiplication by saying that the *set of all integer forms a group under algebraic addition* (and likewise in other cases). We note in passing that the identity element e and the inverse element for a given $a \in G$, are both unique in this example. This will later be shown to be always valid.
2. The set of all rotations about the centre that leaves an equilateral triangle in the x–y plane invariant.
   There are three such operations:

   $R_1 =$ anticlockwise rotation about the z-axis through 120°
   $R_2 =$ anticlockwise rotation about the z-axis through 240°
   $R_3 =$ rotation about the z-axis through 360°

   We note that the multiplication law $\circ$ is the physical operations of rotation taken in succession. We can verify the group properties easily with R_3 as identity and $R_1^{-1} = R_2$.
   A more abstract way of formulating this group is given below.
3. Let $G = (a, a^2, a^3 \equiv e)$.
   Here the elements of G are powers of one group element a, with the cube power identified with e, the identity element. Evidently, the group axioms are satisfied:

   (a) Each element is expressible as a^m with some m and $a^m \circ a^n = a^p$ with $p = (m + n)$ modulo 3 (which means $(m + n - p)$ is a multiple of 3, with $0 \le p \le 2$) also belongs to G.
   (b) Associativity of $\circ$ is evident.

(c) Identity element has been specified.
(d) Inverses of the elements are $(a)^{-1} = a^2$, $(a^2)^{-1} = a$, $(a^3)^{-1} = e$, as is easily verified.

This group is called the *cyclic group* of order 3 and is denoted by C_3.
A realization of this group is obtained when we identify $a \equiv$ the operation of anticlockwise rotation by 120° about the z-axis.
Then, a^3 = the operation of rotation by 360° about the z-axis = e, the identity element. This has been given above.

4. Set of all rational fractions form a group under algebraic multiplication. $S = \{\frac{p}{q};\ p \text{ and } q(\neq 0) \text{ are integers having no common divisor}\}$.
We note here that $e = \frac{1}{1} = 1$ and $\left(\frac{p}{q}\right)^{-1} = \frac{q}{p}$.
In each of the examples above, we note the uniqueness of e and of the inverse of a given element.
5. Consider the following set: $G_6 = \{e, a, a^2, b, ba, ba^2\}$ with $ab = ba^2$ and $a^3 = b^2 = e$. This set forms a group. (The proof is left as an exercise.)

For the purpose of clarity, we add some examples of sets together with 'laws of multiplication' which do *not* form groups.

1. Consider $S \equiv$ the set of all positive integers and o stands for addition of numbers. This set does not constitute a group under addition since the identity element is not found in S. Augmenting S to include the element 'zero' does not still make it a group under addition since (while now an identity element belongs to S) the inverse element of any positive integer does not belong to S.
2. In a similar manner the set of all integers does not from a group under multiplication. This is so because the inverse of an element ($\neq 1$) is not within this set.

More examples of groups will follow in the course of this chapter.

Solved problem 7.1 Show that the set of all unitary unimodular $(N \times N)$ matrices forms a group (unimodular implies that the determinant of the matrix is +1) under matrix multiplication. (This group is called the $SU(N)$ group). Is this a finite group?

Solution: Let S be the set of all unitary unimodular $(N \times N)$ matrices.

1. Now, $A \in S, B \in S \Rightarrow A^\dagger A = I = B^\dagger B$; $\det A = 1 = \det B$ and $(A \circ B) = AB$. Now,

$$(AB)^\dagger (AB) = B^\dagger A^\dagger AB = B^\dagger IB = B^\dagger B = I.$$

So (AB) is a unitary matrix. Further,

$$\det(AB) = \det A \det B = 1.$$

Hence AB is itself a unitary unimodular $(N \times N)$ matrix. Thus, $A \in S, B \in S \Rightarrow AB \in S$.

2. Matrix multiplication is known to be associative.

$$A \circ (B \circ C) = ABC = (A \circ B) \circ C$$

3. $I_{N \times N}$ is the identity element in S; since for every $A \in S$, $AI = A$.
4. Given $A \in S$, $A^\dagger = A^{-1}$ is also a unitary unimodular matrix since

$$(A^\dagger)^\dagger A^\dagger = AA^\dagger = I.$$

[For (finite-dimensional) square matrices, $AB = I \Rightarrow BA = I$], and

$$\det A^\dagger = \det A^{-1} = (\det A)^{-1} = 1.$$

Thus there exists $A^{-1} \in S$ such that $AA^{-1} = I$.

Further, there are an infinite number of 2×2 unitary unimodular matrices. (Prove!) Hence the group is infinite.

Comment: In passing, we note that this is a very important group in physics with applications to angular momentum ($SU(2)$), elementary particle classification ($SU(3)$), theories of fundamental interactions ($SU(3)$, $SU(2)$) and their extensions.

7.2.2 Uniqueness of the identity element and the inverse

When[40] we discussed several examples earlier, we drew attention to the fact that in each case we found (without assumption) that each example had a unique identity element and a unique inverse for each $a \in G$. While defining a group we had *not* demanded that the identity element e and inverse of any element $a \in G$ be unique. It turns out that the set of four axioms themselves imply that that is the case so that it is not required to demand this in the definition. This is similar in spirit to the definition of a linear vector space where one does not have to require the uniqueness of the null vector and the 'reversed' vector.

(The proofs of these statements, to follow, are simple exercises in rudimentary 'group theory thinking'. We note that most of the group theory that is relevant to physics does not use much of any other familiar branch of mathematics, except elementary theory of matrices. Thus, group theory evolves 'its own way of thinking' from scratch so to speak. The following proofs are an introduction to it).

[40] This section can be left out of first reading but the results in it should be noted.

1. **A right inverse is also a left inverse**
   Let e be a right identity element of G. For any $a \in G$, we have,
   $$a^{-1}a = a^{-1}ae = a^{-1}aa^{-1}(a^{-1})^{-1} = a^{-1}(aa^{-1})(a^{-1})^{-1}$$
   $$= a^{-1}e(a^{-1})^{-1} = e.$$
   We note that in the second step we have chosen to express e as: $e = a^{-1}(a^{-1})^{-1}$ in terms of the element a^{-1} and its inverse $(a^{-1})^{-1}$. Both of these are known to belong to G from axiom (4) in the definition of a group.
   The above relation shows that a^{-1}, *a right* inverse, is also *a left* inverse.
2. **A right identity is also a left identity**
   For any $a \in G$, $ea = (aa^{-1})a = a(a^{-1}a) = ae = a$.
   Thus, e, a *right* identity, is also *a left* identity.
3. **Law of cancellation**
   $ab = ac$ for some $a \in G$ implies $b = c$. This is seen by pre-multiplying by a^{-1} which also is a left inverse.
   $$a^{-1}ab = a^{-1}ac \Rightarrow eb = ec$$
   As e is a left identity this implies $b = c$.
4. **The uniqueness of the identity element**
   Let, if possible, there be another (right) identity element e'. Then for $a \in G$,
   $$ae = a = ae'.$$
   But $ae = ae' \Rightarrow e = e'$ by (3) above. Hence the right identity is unique. This is also a left identity. In a similar manner we can show that e is the *unique left identity*. This proves the uniqueness of the identity element.
5. **The uniqueness of the inverse element**
   Consider an $a \in G$. Let, if possible, there be two right inverse elements $b, c \in G$ for a given $a \in G$. Then,
   $$ab = e = ac.$$
   By (3) above, however, this implies $b = c$. This proves the uniqueness of the right inverse. The right inverse has been shown to be a left inverse and it can be shown to be the unique left inverse along the same lines.

   Additional observations
6. We further note that $(a^{-1})^{-1} = a$.
   This follows since
   $$(a^{-1})(a^{-1})^{-1} = e = a^{-1}a.$$
   We pre-multiply by a to obtain the result.

7. The inverse of a product of elements is given by

$$(ab\ldots c)^{-1} = c^{-1}\ldots b^{-1}a^{-1}$$

since

$$\begin{aligned}(ab\ldots c)^{-1}(ab\ldots c) &= c^{-1}\ldots b^{-1}a^{-1}ab\ldots c\\ &= c^{-1}\ldots b^{-1}eb\ldots c\\ &= c^{-1}\ldots b^{-1}b\ldots c\\ &= c^{-1}\ldots e\ldots c = e.\end{aligned}$$

7.2.3 Further definitions

Finite/infinite groups and order of a finite group: If the number of elements of a group G is finite (say N), G is called a finite group of the order N. Otherwise G is an infinite group.

Examples (1) and (4) in the section 7.2.1 are infinite groups. Examples (2), (3) have a finite group of order 3. Example (5) is a group of order 6.

Abelian and nonabelian Groups: If, for *all* $a, b \in G$, $ab = ba$, then G is an *abelian group*. If a group is not abelian, it is called a *nonabelian* group.

For a nonabelian group G there exists at least one pair of elements $a, b \in G$ such that $ab \neq ba$. (Note that a nonabelian group *can* have pairs of elements which commute.). The first four examples in Section 7.2 are of abelian groups while the last one is of a nonabelian group.

Examples An example of a nonabelian group is the group of 2×2 real non-singular matrices under the law of matrix multiplication. As is known, any arbitrary two 2×2 matrices need not commute. The example of $SU(N)$ introduced earlier in Section 7.2.1 is also a nonabelian group.

Another example in a physical context would be the group of rotations in 3-dimensions. It is known that two finite rotations need not generally commute.

Example (5) of Section 7.2.1 is another example of a nonabelian group.

Below, we mention a few simple results that will follow during the course of the discussion in future sections.

1. A group whose order is 5 or less is necessarily abelian. The smallest nonabelian group is G_6 of Example (5) in the Section 7.2.1.
2. A group whose order is prime is necessarily abelian (see Exercise 7.7).

7.2.4 Structure of the group multiplication table for a finite group

Let $G = \{e \equiv a_0, a_1, a_2, \ldots, a_{n-1}\}$ be a (finite) group of order n. We construct an $(n \times n)$ table in which each row and each column is labelled by a group element in the above order.

In the row labelled by a_m and the column labelled by a_p, we write the result for the product $a_m \circ a_p$ (note: this may not equal $a_p \circ a_m$) [see table below]. A rudimental knowledge of group G which consists of the set of n elements together with the law of multiplication, is completely obtained if this table is given (and vice versa). A convenient way to represent all the basic knowledge of a group (especially of a small order i.e., small n) is, therefore, to give the multiplication table.

Table 7.1 *A typical multiplication table*

	$a_0 = e$	a_1	a_2	$\cdots$	a_p	$\cdots$	a_{n-1}
e	e	a_1	a_2	$\cdots$	a_p	$\cdots$	a_{n-1}
a_1	a_1						
a_2	a_2						
$\cdots$	$\cdots$						
a_m	a_m				$a_m \circ a_p$		
$\cdots\cdots$	$\cdots\cdots$						
a_{n-1}	a_{n-1}						

Properties of the group multiplication table

We note some of the obvious properties of the above table:

1. The first row consists of the group elements in the sequential order ($e \circ a_p = a_p$). The same applies to the first column ($a_p \circ e = a_p$).
2. Each element appears exactly once in each row. Suppose, some element a_q appears (at least) twice in, say, the mth row[41] and the rth and the sth columns for some r, s $(0 \leq r, s \leq n-1)$. Then $a_m \circ a_r = a_q = a_m \circ a_s$. Pre-multiplying by a_m^{-1}, we obtain $a_p = a_s$. But the elements a_p and a_s are distinct by the definition of a group G of order n. Hence each element in the mth row must be distinct. Furthermore, as there are n elements in row, each must appear exactly once.
3. In a similar manner, each element appears exactly once in each column.
4. Consequently, each row is permutation of $(a_0, a_1, \ldots a_{n-1})$. And distinct rows are *distinct* permutations. (Prove!)
5. The above observation also applies to columns.

Examples The following two tables represent multiplication tables for two groups, of order 3 and the other of order 4 respectively.

[41] We actually mean the row labelled as a_m.

Table 7. 2

	e	a	a^2
e	e	a	a^2
a	a	a^2	e
a^2	a^2	e	a

Table 7. 3

	e	a	b	c
e	e	a	b	c
a	a	e	c	b
b	b	c	e	a
c	c	b	a	e

Comment: We note that the sequence of elements in a group G is by no means fixed. If we change this sequence, the rows and simultaneously the columns will get shuffled and we will have a multiplication table with a different *appearance* for the same group but having the same content.

Solved problem 7.2 Using *only* the multiplication table properties, construct the group multiplication table for a group of order 3 and show that it coincides with Table 7.2 above.

Solution: Let $G = \{e, a, b\}$ be the group. Consider filling up, as far as possible, the multiplication table making use of the properties listed above.

	e	a	b
e	e	a	b
a	a	?	??
b	b	???	????

The first row and the column are fixed. The entry (?) cannot be e because that would require (??) to be b. Then two elements in the last column would coincide. Thus, the entry (?) must be b. That leaves (??) to be e. Now, the columns 2 and 3 each must have all distinct elements. This requires (???) to be e and (????) to be a. This in fact then coincides with Table 7.2 above with $b = a^2$.

Exercise: Write down all possible multiplication tables for a group of order 4 and hence show that they are abelian.

7.2.5 Cyclic groups and order of an element

Consider an element a together with a definition of the multiplication law so that a^r is defined for $r = 2, 3, \ldots$. Suppose $a^m \equiv e$ and $a^r \neq e,\ r < m$. Then the set $G = \{e, a, a^2, \ldots a^{m-1}\}$

forms a group. (See Exercise 7.3). This group is called a cyclic group of order m, denoted usually by C_m. A cyclic groups is always abelian (see Exercise 7.3). We further note that the inverse of the element a^p with $0 < p \leq m$ is a^{m-p} since $a^p a^{m-p} = a^m = e$.

Order of an element of a finite group G

Let $a \in G$, a finite group of order n. Now consider the powers of a up to the nth power: $a, a^2, a^r, \ldots a^n$. We shall show that at least one of $\{a^r; 1 \leq r \leq n\}$ must be equal to e. Consider the following $(n+1)$ elements $\{e, a, a^2, \ldots a^r, \ldots, a^n\}$. Now, G is of the order n and hence cannot have more than n distinct elements. Hence at least two of these must be equal. Then either $e = a^p (1 \leq p \leq n)$, in which case the result in proven, and/or $a^r = a^s$ with $r < s \leq n$. Then $e = (a^r)^{-1} a^r = (a^r)^{-1} a^s = (a^{-1})^r a^s = a^{s-r}$. (Why?) Thus, a power $(s-r)$ (with $1 \leq s-r < n$) of a equals e. It thus follows that there exists an integer p, $1 \leq p \leq n$ such that $a^p = e$. The smallest value of p is called the *order* of the element a, and is denoted by $O(a)$. As shown, $O(a) \leq O(G)$.

Order of an element

Let $a \in G$ (not necessarily a finite group). If there exists a positive integer p such that $a^p = e$ and $a^r \neq e$, $1 \leq r < p$; then the element a has order p. If there is no such integer, then the order of a is infinite. As remarked earlier, the order of an element of a *finite* group is always finite and $\leq O(G)$. The order of any element of G other than the identity is necessarily greater than 1 (see Exercise 7.2).

A result

It can be shown that (see Exercise 7.6) that the order p of an element a belonging to a group of order n is a divisor of n.

This can also be seen as a corollary to the Lagrange's theorem in Section 7.3.2.

Examples

1. Let G be a group of all rotations about the z-axis. Let $R_n \in G$ be a rotation through $2\pi/n$. Then the lowest power p of R_n that equals the identity is n:$R_n^n = e$ as this represents a rotation through 2π. Thus, R_n has order n. On the other hand, let $R \in G$ be a rotation through 1 radian. Then the order of R is not finite since for any integer N, $R^N \neq e$. This is so since R^N represents a rotation through N radians which can never equal a multiple of 2π, an irrational number.
2. Let $G =$ the group of all unitary unimodular 2×2 matrices ($SU(2)$). The order of $\begin{bmatrix} 1 & 0 \\ 0 & -1 \end{bmatrix}$ is two. Order of $\begin{bmatrix} e^{i2\pi/3} & 0 \\ 0 & e^{-i2\pi/3} \end{bmatrix}$ is three. There are many elements of G of an infinite order.

Solved problem 7.3 Let $G = \{e, a, b, c\}$.

(a) What are the possible orders of each of the elements of G other than the identity?
(b) If each of the elements a, b, c have the same order, what can it be?
(c) If the order of a, b, c are not all identical, what can you say about these orders?

Solution:

1. G is a group of order four. Hence the possible the order of a, b and c are two or four.
2. Let a have order 4. Then $a^4 = e$. Then G must be identical to $C_4 \equiv \{e, a, a^2, a^3\}$. Then the one of b and c must be a^2. Then the the order of this element must be two ($(a^2)^2 = e$). Thus, all of (a, b, c) cannot have an order 4.
   On the other hand, each of (a, b, c) can have order two. Then $a^2 = b^2 = c^2 = e$. This leads Table 7.3.
3. If the orders of a, b, c are not all identical, then one of these (say for a) must be 4. Then $a^2 \equiv b$ has order 2 and $a^3 = c$ has order 4. (It cannot be two since $(a^3)^2 = a^6 = a^2 = b \neq e$.) Thus two of (a, b, c) have order four and one has order) two.

7.3 Subgroups and Direct Product of Groups

7.3.1 Subgroups

A subgroup of a group G is essentially a subset of G that forms a group by itself. This must be with the *same* law of group multiplication.

Definition: Consider a group G. Let H be a subset of elements of G such that

(i) for every $a, b \in H, a \circ b \in H$ (under the same law of group multiplication law $\circ$ as that of G).
(ii) for every $a \in H, a^{-1}$ also belongs to H.

Then H is called a subgroup of G.

Comments:

(a) We note that the two requirements above are just the two of the group axioms that are necessary for H to be a group.
(b) We do not need to specify here the associativity of group multiplication $\circ$, since associativity of the group multiplication for the elements of $H \subset G$ is already given to us since G is a group.

(c) We do not need to additionally specify the existence of an identity e in H. This is so because the requirements (i) and (ii) automatically imply that given $a \in H$, $aa^{-1} = e \in H$.

(d) This is regarding the comments (b) and (c) above. In this connection one may recall an analogous situation that arises while defining a subspace of a linear vector space (see e.g., Chapter 4 of *Mathematical Physics: The Basics*).

(e) For finite groups and also for infinite groups all whose elements are of *finite order*, only condition (i) is needed. To see this, consider any element $a \in G$ with $O(a) = n$. Then $a^{-1} = a^{n-1} = (a \circ a \circ \ldots a)$, which is known to belong to H by (i). Thus condition (ii) becomes redundant in such cases.

An alternate definition of a subgroup

It is possible to combine conditions (i) and (ii) in the definition of a subgroup in the form of a *single* condition (iii) given below.
(iii) For every $a, b \in H$, ab^{-1} also belongs to H. (The proof of this is left as an exercise: see Exercise 7.10). Verification of whether a subset $H \subset G$ is a subgroup of G is often facilitated by this latter condition.

Trivial and proper subgroups

The entire group G and the one-element set $\{e\}$ are always subgroups of G. These are called the *trivial* or *improper* subgroups. All other subgroups of G are called *proper* subgroups.

An invariant subgroup

A subgroup H of G is called an *invariant* subgroup of G if, for every $h \in H$ and $\forall\, g \in G$, $ghg^{-1} \in H$.

Thus, a subgroup H is left invariant by every 'operation' $g \in G$ in the sense that $gHg^{-1} = H$.

Comments:

(a) If G is an abelian group, every subgroup H of G is automatically an invariant subgroup since for every $g \in G$ and each $h \in H$, $ghg^{-1} = h \in H$.

(b) $\{e\}$ and G are always invariant subgroups of G. (Why?)

(c) For a finite group G, the possible order of a subgroup H of G is restricted. Lagrange's theorem states that $O(H)$ is a divisor of $O(G)$. This will be proved subsequently in Section 7.3.2.

Examples

1. Consider $G = C_4 = \{e, a, a^2, a^3\}$, the cyclic group of order 4. Let $H = \{e, a^2\}$. Then H is (i) a proper and (ii) and invariant subgroup of G. (Verify!) That it is an invariant subgroup of G follows from the abelian nature of G. Note that $(a^2)^{-1} = a^2$.

2. Let G be the set of all integers that forms a group under addition. Let E = set of all even integers. Then E is a proper subgroup of G. Moreover, as G is abelian, E is an invariant subgroup.
3. Let G be the set of all proper rotations about a point. Let $\Re_z$ be the set of all rotations about the z-axis. Then, $\Re_z$ is a proper subgroup but not an *invariant* subgroup of G. This is because finite rotations do not generally commute.

Solved problem 7.4 Show that the set G of all 2×2 complex matrices forms a group under matrix addition. Is it abelian? Show further that the set H of all hermitian 2×2 matrices is a proper subgroup of G. Is it an invariant subgroup? Can you suggest some more proper subgroups of G?

Solution: Let $A, B \in G$. Then $A + B$ is a complex 2×2 matrix and hence belongs to G. The matrix addition is associative. Further, the null matrix $\left[\begin{smallmatrix} 0 & 0 \\ 0 & 0 \end{smallmatrix}\right] \in G$ is the identity element and $(-A) \in G$ is the inverse element of A.

The group is abelian since $A + B = B + A$ for all $A, B \in G$.

To show that the set H of all hermitian 2×2 matrices is a proper subgroup of G, we note that for every $A, B \in H$, $A \circ B^{-1} \in H$. This is so since A and B are hermitian, so is $A \circ B^{-1}$ which is obtained as the matrix $A + (-B) = A - B$ which is hermitian. This implies that H is a subgroup of G. Further, G contains in it antihermitian matrices while H does not, so that H is a proper subgroup of G.

Further since G is an abelian group, every subgroup of G, and thus H, is an invariant subgroup.

Another proper subgroup of G is the set of all 2×2 antihermitian matrices. The set of all real 2×2 matrices is also another proper subgroup.

7.3.2 Coset decomposition

Consider a group G (not necessarily finite). Let $H = \{h_1, h_2, \ldots, h_n, \ldots\}$ be a proper subgroup of G. Then, there exists an element $a \in G$ such that $a \notin H$. We now consider the set of elements $\{ah_1, ah_2, \ldots, ah_n, \ldots\} \subset G$. We note that all these elements are distinct, for if we had $ah_r = ah_s$, this would imply $h_r = h_s$ which is false, as these are distinct elements of H. Moreover, none of these elements, ah_i, belong to H. For if we had, $ah_r \in H$, that would imply $ah_r = h_s$ for some s, which in turn would imply $a = h_s h_r^{-1} \in H$ since H is a subgroup. But $a \notin H$ by assumption. We shall denote the set $\{ah_1, ah_2, \ldots, ah_n, \ldots\}$ as aH. Now, consider the set $G - H - aH$. If this set is empty, we can write

$$G = H + aH.$$

On the other hand, if $G - H - aH \neq \emptyset$, then there exists an element $b \in G$ such that $b \notin H$ and $b \notin aH$. Now starting with b, we form the set $bH \equiv \{bh_1, bh_2, \ldots, bh_n, \ldots\} \subset G$. It is easy to show (as done previously) that all the elements of bH are distinct and are moreover distinct from those of both H and aH. Again, if $G - H - aH - bH = \emptyset$, then we can write

$$G = H + aH + bH.$$

If not, we continue this process further. We thus arrive at

$$G = H + aH + bH + \cdots + uH + \cdots \tag{7.1}$$

If G is a finite group, (and hence H is finite), the above process must end in a finite number of steps, say r. In that case, we have

$$O(G) = rO(H), \tag{7.2}$$

i.e., *the order of a subgroup H is a divisor of the order of G*. This is called the *Lagrange's theorem*.

Irrespective of whether G is a finite group, the above decomposition (7.1) of G is called a *coset decomposition* and the sets $aH, bH, \ldots$ are called the *left cosets*. A similar decomposition can be arrived at in terms of the *right cosets*:

$$G = H + Ha' + Hb' + \cdots + Hu' + \cdots \tag{7.3}$$

Examples

1. Let G be the group of order 6 introduced in the Section 7.2: $G = \{e, a, a^2, b, ba, ba^2\}$ with $ab = ba^2$. Proper subgroups of G are $H_1 = \{e, a, a^2\}$ and $H_2 = \{e, b\}$. We note that the orders of H_1 and H_2 are 3 and 2 respectively and are divisors of $O(G) = 6$. A left coset decomposition in terms of the cosets of H_1 is $G = H_1 + bH_1 = \{e, a, a^2\} + \{b, ba, ba^2\}$ and a right coset decomposition in terms of the cosets of H_2 is $G = H_2 + H_2a + H_2a^2 = \{e, b\} + \{a, ba\} + \{a^2, ba^2\}$.
2. Let G be the group (denoted by $O(3)$) of all rotations (proper as well as improper) in 3-dimensions and let H be the subgroup of all proper rotations (denoted by $SO(3)$). Let I stand for the inversion. Then, $G = H + IH = H + HI$. The set of all improper rotations is not a subgroup but a coset $IH = HI$.
3. Let G be the set of all integers under addition and let E be the subgroup of all even integers. Then, G has a left coset decomposition:

   $$G = E + 1 \circ E$$

   and a right coset decomposition:

   $$G = E + E \circ 1.$$

   If, on the other hand, Q be the subgroup of all multiples of 5, then,

   $$G = Q + 1 \circ Q + 2 \circ Q + 3 \circ Q + 4 \circ Q$$

   is its left coset decomposition.

Corollary to the Lagrange's theorem: Let $a \in G$, a finite group. Let $O(a) = p \leq O(G)$. then $\mathcal{C}_p = \{e, a, a^2, \ldots, a^{p-1}\} \subset G$ is a subgroup of G of order p. Hence $O(a) = p$ must be a division of $O(G)$. Hence, should $O(G)$ be prime, $p = O(G)$ and $G \equiv \mathcal{C}_p$, which is cyclic and abelian.

7.3.3 Direct product of groups

Let $G_1 = \{e_1, a_1, b_1, \ldots\}$ and $G_2 = \{e_2, a_2, b_2, \ldots\}$ be two groups. We consider the set $G \equiv G_1 \otimes G_2$. A typical element of the set $G_1 \otimes G_2$ is an ordered pair (g_1, g_2) with $g_1 \in G_1$ and $g_2 \in G_2$. On the set $G_1 \otimes G_2$, we define an operation of multiplication as

$$(g_1, g_2) \circ (g_1', g_2') = (g_1 \circ g_1', g_2 \circ g_2')$$

which is an element of G since $g_1 \circ g_1' \in G_1$ and $g_2 \circ g_2' \in G_2$. Then we shall show that G forms a group under this multiplication law. We note:

(i) $\forall g_1, g_1' \in G_1$ and $g_2, g_2' \in G_2$, we have $(g_1, g_2) \circ (g_1', g_2') = (g_1 \circ g_1', g_2 \circ g_2') \equiv (g_1'', g_2'') \in G$.

(ii) The group multiplication law on G is associative. This follows from the associativity of the group multiplications on G_1 and G_2.

(iii) (e_1, e_2) serves as the identity element on G since

$$(e_1, e_2) \circ (g_1, g_2) = (e_1 g_1, e_2 g_2) = (g_1, g_2).$$

(iv) (g_1^{-1}, g_2^{-1}) is the inverse of (g_1, g_2) for any $(g_1, g_2) \in G$

The proof of statements (ii) and (iv) is left as an exercise.

Comments:

(a) If G_1 and G_2 are groups of orders m_1 and m_2 respectively, then G is of order $m_1 m_2$.

(b) G has as its invariant subgroups $G_1' = \{(e_1, e_2), (a_1, e_2), (b_1, e_2), \ldots\}$ and $G_2' = \{(e_1, e_2), (e_1, a_2), (e_1, b_2), \ldots\}$. G_1' 'isomorphic' to $G_1 (G_1' \approx G_1)$ and $G_2' \approx G_2$.[42] (See Exercise 7.12).

Example Consider $G_1 = \{e_1, a\}$ and $G_2 = \{e_2, b\}$. Then $G_1 \otimes G_2 \equiv \{(e_1, e_2), (a, e_2), (e_1, b), (a, b)\}$. It is easy to verify that $G_1 \otimes G_2 \approx \mathcal{V} \equiv \{e, A, B, C\}$ with $A^2 = B^2 = C^2 = e$. $\mathcal{V}$ follows the multiplication rules given in Table 7.3.

[42]Two groups G and G' are said to be isomorphic iff there is a one-to-one correspondence between their elements preserved under group multiplication. We then write $G \approx G'$.

7.4 Matrix Representations of a Group

In this section we shall deal with the matrix representations of a group, generally a finite one. The scope of the book allows for a brief discussion, and often without proof[43].

7.4.1 Definitions

Let $G = \{a, b, c, \dots\}$ be a group. Let us associate an $N \times N$ matrix $D(g)$ with each element $g \in G$. (The matrices corresponding to distinct elements of G need not necessarily be distinct). Now, suppose these matrices satisfy $D(a)D(b) = D(c)$ whenever $a \circ b = c$ for all $a, b \in G$. Then, the set[44] of $N \times N$ matrices $D(G) \equiv \{D(a), D(b), D(c), \dots\}$ is called an $N \times N$ matrix representation of G.

If, in a representation, each element of G is associated with a *distinct* $N \times N$ matrix, then the representation is called a *faithful* representation. If this is not the case, the representation is called an *unfaithful* representation.

Examples

1. Consider the cyclic group $C_3 = \{e, a, a^2\}$. Then, the following set of 2×2 matrices constitute a representation of C_3.

$$D(e) = I \tag{7.4}$$

$$D(a) = \frac{1}{2}\begin{bmatrix} e^{\frac{2\pi i}{3}} + 1 & e^{\frac{2\pi i}{3}} - 1 \\ e^{\frac{2\pi i}{3}} - 1 & e^{\frac{2\pi i}{3}} + 1 \end{bmatrix} \tag{7.5}$$

$$D(a^2) = [D(a)]^2 \tag{7.6}$$

It can be easily be verified that $(D(a))^3 = I = D(e)$. Thus, the set $\{D(e), D(a), D(a^2)\}$ constitutes a 2×2 representation of G.

2. Consider again $G = C_3$ given above. Let

$$D^{(1)}(e) = 1; \quad D^{(1)}(a) = e^{\frac{2\pi i}{3}}; \quad D^{(1)}(a^2) = e^{\frac{4\pi i}{3}}.$$

This set constitutes a 1×1 representation of G.
Also, the set

$$D^{(2)}(e) = 1 = D^{(2)}(a) = D^{(2)}(a^2)$$

constitutes a 1×1 representation of G called the *trivial* representation.

---

[43]For fuller treatment, one can refer to *Group Theory* by Hamermesh.

[44]While we shall use the notation of a set, we note that a matrix in the set may be occurring more than once in the set.

3. Consider the group $G_6 = \{e, a, a^2, b, ba, ba^2\}$. The following set of matrices constitutes a 2×2 representation.

$$D(e) = \begin{bmatrix} 1 & 0 \\ 0 & 1 \end{bmatrix} \quad D(a) = \begin{bmatrix} \omega & 0 \\ 0 & \omega^2 \end{bmatrix} \quad D(a^2) = \begin{bmatrix} \omega^2 & 0 \\ 0 & \omega \end{bmatrix}$$

$$D(b) = \begin{bmatrix} 0 & 1 \\ 1 & 0 \end{bmatrix} \quad D(ba) = \begin{bmatrix} 0 & \omega^2 \\ \omega & 0 \end{bmatrix} \quad D(ba^2) = \begin{bmatrix} 0 & \omega \\ \omega^2 & 0 \end{bmatrix}$$

$$\omega \equiv \exp(2\pi i/3) \quad 1 + \omega + \omega^2 = 0. \tag{7.7}$$

Comments:

(a) In the above examples, the representations $D(G)$ and $D^{(1)}(G)$ associate a *distinct* matrix with each element of G. Hence they are faithful representations. On the other hand, $D^{(2)}(G)$ associates the *same* matrix with more than one element of G (here all three). Hence it constitutes an unfaithful representation.

(b) We note that we can always associate the $N \times N$ identity matrix with *all* the elements of a group G. This will constitute an unfaithful representation of G.

7.4.2 Equivalent representations

Consider a group $G = \{a, b, c, \dots\}$ together with an $N \times N$ representation $D(G) = \{D(a), D(b), D(c), \dots\}$. Consider another set of matrices, $D'(G)$ obtained by applying the *same* similarity transformation S to *all* the representation matrices:

$$D'(G) = \{D'(a), D'(b), D'(c), \dots\}$$

with $D'(g) = SD(g)S^{-1}\ \forall\, g \in G$. We shall show that $D'(G)$ is also an $N \times N$ representation of G. To see this, consider an arbitrary pair of elements $(g, f) \in G$. Then, since $D(G)$ is a representation of G, we know that $g \circ f = h$ implies $D(g)D(f) = D(h)$. We can rewrite this equation as $SD(g)S^{-1}SD(f)S^{-1} = SD(h)S^{-1}$, i.e., $D'(g)D'(f) = D'(h)$. This implies that $D'(G)$ is a set of $N \times N$ matrices in correspondence to the elements of G such that for each pair $(g, f) \in G$, $g \circ f = h$ implies $D'(g)D'(f) = D'(h)$. Thus, $D'(G)$ is also an $N \times N$ representation of G. We note that $D(G)$ is also related to $D'(G)$ by a similarity transformation: $D(g) = S^{-1}D'(g){S^{-1}}^{-1}$.

Definitions: Two representations $D'(G)$ and $D(G)$ related to each other by a similarity transformation are said to be *equivalent*.

Any pair of representations $D'(G)$ and $D(G)$ which are not equivalent are said to be *inequivalent*.

Example It is easy to verify that the 2 × 2 representation $D(G)$ of C_3 considered in the previous subsection is equivalent to the following representation $D'(G)$, where

$$D'(g) = SD(g)S^{-1} \qquad \forall g \in G,$$

with,

$$S = \frac{1}{2}\begin{bmatrix} 1 & 1 \\ 1 & -1 \end{bmatrix}. \tag{7.8}$$

We then have the following explicit matrices:

$$D'(e) = D(e) = I$$

$$D'(a) = SD(a)S^{-1} = \begin{bmatrix} e^{\frac{2\pi i}{3}} & 0 \\ 0 & 1 \end{bmatrix}$$

$$D'(a^2) = SD(a^2)S^{-1} = \begin{bmatrix} e^{\frac{4\pi i}{3}} & 0 \\ 0 & 1 \end{bmatrix}$$

We note that the all the matrices of the new representation $D'(G)$ are diagonal. We also note that the diagonal entries are nothing but those from the two one-dimensional representations $D^{(1)}(G)$ and $D^{(2)}(G)$ discussed in the second example of the previous subsection.

Comments:

(a) Given an $N \times N$ representation $D(G)$, there are an infinite number of equivalent $N \times N$ representations of G. They do not contain 'new information' in the sense that once one member of the set of equivalent representations is known, all the rest are, in principle, known. In fact, there is a way of characterising a representation *independent of equivalence* in terms of what are called 'characters', which we shall later touch upon briefly.

(b) Evidently, if all the *inequivalent* representations of G are known, all the representations of G would be known.

(c) In the previous example we saw that the two-dimensional representation $D(G)$ could be constructed from the one-dimensional representations $D^{(1)}(G)$ and $D^{(2)}(G)$ by first (i) putting them together in the form of diagonal matrices and then (ii) applying a similarity transformation S^{-1}. This shows that a representation (here $D(G)$) may be equivalent to the one (here $D'(G)$) that can be 'reduced to' other smaller dimensional representations (here $D^{(1)}(G)$ and $D^{(2)}(G)$). In the next section we shall discuss this question of 'reducibility' of a representation.

(d) We shall mention in passing a result regarding finite dimensional representations of finite groups. It can be shown that they are always equivalent to unitary representations (i.e., one in which *all* representation matrices are unitary).

7.4.3 Reducible and irreducible representations

Consider a group $G = \{a, b, c, \dots\}$ and an $N \times N$ representation of G:

$$D(G) = \{D(a), D(b), D(c), \dots\}.$$

If there exists a similarity transformation S such that each of the matrices of the transformed representation $D'(G) = SD(G)S^{-1}$ has the *block diagonal form*

$$\begin{bmatrix} A_1 & 0 & 0 & & 0 \\ 0 & A_2 & 0 & & 0 \\ 0 & 0 & A_3 & & 0 \\ .. & .. & .. & .. & . \\ 0 & 0 & 0 & .. & A_n \end{bmatrix}, \tag{7.9}$$

where A_r, $r = 1, 2, \dots, n(> 1)$ are square matrices of dimension d_r such that $\sum_{r=1}^{r=n} d_r = N$, then we say the the representation is (fully) reducible.

On the other hand, for a finite G, if this is not possible for *any* S, then the representation is called irreducible[45].

Comments:

(a) It is easy to show that the set $D_r(G)$ of the $d_r \times d_r$ blocks, viz., $\{A_r(a) : a \in G\}$ constitutes a $d_r \times d_r$ representation of G. For a fully reducible representation, a similarity transformation exists for which each of the blocks is an irreducible representation of G (see Exercise 7.13).

(b) As commented in a subsequent subsection 7.4.5, there is an upper bound on the dimension of an irreducible representation for a given finite group. Then, any representation of dimensions higher than this limit is necessarily reducible. For an abelian group, it will turn out that every representation of dimension ≥ 2 is necessarily reducible.

(c) It is possible to develop a criterion to test whether a given representation is reducible *without* ever searching for a similarity transformation. It is based on what are called 'characters' introduced briefly in the next subsection. It is also possible to determine the possible dimensions of irreducible representations that can enter in the block-diagonal structure.

(d) The problem of finding all possible representations is further reduced to finding all possible *inequivalent irreducible* representations. As we have remarked in Section 7.4.5, for a given finite group G, there are a limited number of these.

[45] For finite groups, *all* finite dimensional representations are either fully reducible or irreducible. See e.g., *Group theory* by M. Hamermesh.

Examples We have already given examples by way of introduction. We summarise these and add further.

1. The 2×2 representation of example 1 of Section 7.4.1 is a fully reducible representation. As shown earlier, a similarity transformation given by the matrix S of (7.8) transforms it into a block-diagonal form with two 1×1 blocks.
2. Consider the abelian group $V \equiv \{e, a, b, c\}$ with $a^2 = b^2 = c^2 = e$. Consider the 4×4 representation, called the regular representation:

$$D(e) = \begin{bmatrix} 1 & 0 & 0 & 0 \\ 0 & 1 & 0 & 0 \\ 0 & 0 & 1 & 0 \\ 0 & 0 & 0 & 1 \end{bmatrix} \quad D(a) = \begin{bmatrix} 0 & 1 & 0 & 0 \\ 1 & 0 & 0 & 0 \\ 0 & 0 & 0 & 1 \\ 0 & 0 & 1 & 0 \end{bmatrix}$$

$$D(b) = \begin{bmatrix} 0 & 0 & 1 & 0 \\ 0 & 0 & 0 & 1 \\ 1 & 0 & 0 & 0 \\ 0 & 1 & 0 & 0 \end{bmatrix} \quad D(c) = \begin{bmatrix} 0 & 0 & 0 & 1 \\ 0 & 0 & 1 & 0 \\ 0 & 1 & 0 & 0 \\ 1 & 0 & 0 & 0 \end{bmatrix} \tag{7.10}$$

For an abelian group, every representation of dimension > 1 is reducible. Indeed a similarity transformation

$$S = \frac{1}{2}\begin{bmatrix} 1 & 1 & 1 & 1 \\ 1 & -1 & 1 & -1 \\ 1 & 1 & -1 & -1 \\ 1 & -1 & -1 & 1 \end{bmatrix}$$

renders all $D(g)$, $g \in G$ in diagonal form.

7.4.4 Class structure of a group and characters

In the representation theory for finite groups, the class structure of a group G and the notion of a 'character' proves extremely important.

Definitions: Let $a, b \in G$. If there exists $g \in G$ such that $gag^{-1} = b$, then b is said to be **conjugate** to a and we denote it by $b \sim a$.

If $b \sim a$, then $a = hbh^{-1}$, with $h \equiv g^{-1} \in G$. Thus $a \sim b$. Obviously, $a \sim a$.

We note that an element $a \sim b = gag^{-1}$ has a physical meaning when the group elements are 'operations' (such as rotations, reflections and inversion). $a \sim b$ means that element a 'transforms into' b by the group 'transformation' g. Thus, the elements in the same class are 'similar kind of' operations.

Definition: Let $a \in G$. Then the set of all elements in G conjugate to a is called a **class** in G.

A finite group G is broken up into c classes, $C_1, C_2, \ldots C_c$, and the classes are non-overlapping sets: $C_r \bigcap C_s = \emptyset$ for all $r \neq s$ (see Problem 7.14).

We summarise several obvious results:

(a) The identity element e is a class by itself.

(b) For an abelian group, each class has only one element (since $gag^{-1} = a \, \forall g \in G$). For a finite abelian group, the number of classes 'c' is also the number of elements $g = O(G)$.

(c) For a nonabelian group, there is at least one pair such that $ab \neq ba$ Thus, the element $aba^{-1} \in G$ is conjugate to b but distinct from b. Thus there is at least one class in G having more than one element.

(d) Consequently, for a finite nonabelian group, $c < g$.

Example A nontrivial example of a class structure is possible only for a nonabelian group. For the simplest nonabelian group $G_6 = \{e, a, a^2, b, ba, ba^2\}$, we note:

- $bab^{-1} = bab = bba^2 = a^2 \Rightarrow a^2 \sim a$
- $aba^{-1} = aba^2 = ba^2a^2 = ba \Rightarrow b \sim ba$
- $a^2b(a^2)^{-1} = a^2ba = aba^2a = ab = ba^2 \Rightarrow b \sim ba^2$

There are no further equivalence relations. This implies the following class decomposition:

$$G_6 = \{e\} + \{a, a^2\} + \{b, ba, ba^2\}$$

Definition: Let $D(G) = \{D(a), D(b), D(c), \ldots\}$ be a representation of group $G = \{a, b, c, \ldots\}$. Then the character $\chi(a)$ associated with the element a is defined by

$$\chi(a) = \text{tr}\,[D(a)].$$

The set of characters for the representation $D(G)$ is $\{\chi(a), \chi(b), \chi(c), \ldots\}$.

Characters play a central role in the theory of representations of a finite group. Most operations involving representations can be done using only the characters without ever having to touch representation matrices.

Examples

1. The two-dimensional representation in example 3 of Section 7.4.1 has the following characters.

$$\chi(e) = 2, \chi(a) = \chi(a^2) = -1, \chi(b) = \chi(ba) = \chi(ba^2) = 0$$

2. The two dimensional representation in the example 1 of the same section has

$$D(e) = 2, D(a) = 1 + \omega, D(a^2) = 1 + \omega^2; \omega \equiv \exp(2\pi i/3).$$

3. For any one-dimensional representation, the representation matrices (1×1) and the characters obviously coincide.

We summarise several important properties of characters. These are left as exercises (see Exercise 7.21).

(a) Character is unchanged by a similarity transformation and hence the set of characters is the same for all equivalent representations.
(b) All elements in the same class have the same characters (since $b \sim a \Rightarrow b = gag^{-1}$ for some $g \in G \Rightarrow D(b) = D(g)D(a)D^{-1}(g) \Rightarrow$ tr $[D(b)] =$ tr $[D(a)]$).
(c) For an $N \times N$ representation, $\chi(e) = \text{tr } I = N$.
(d) For a one-dimensional representation, $D(a) = \chi(a)$.
(e) For an $N \times N$ unitary representation, $|\chi(a)| \leq N$. On account of the equivalence of any $N \times N$ representation to a unitary representation, this property holds for *all* $N \times N$ representations.

7.4.5 Summary of results on representations of finite groups

In this section we shall summarise, without proof, the main results on the representations of a finite group G. These results, simple yet powerful in scope, are essential in their applications to spectroscopy of crystals and molecules.

1. A finite group has only a finitely many inequivalent irreducible representations (irreps[46]). In fact, a group of order g has at the most g irreps.
2. The maximum dimensionality of an irrep of a group of order $g(> 1)$ is restricted; in fact, it is bounded by $\sqrt{g-1}$.
3. The number c of classes in G is always less than or equal to g. There is a powerful result that states that the number of inequivalent irreps is equal to c.
4. Let $\mu = 1, 2, \ldots, c$ denote inequivalent irreps and n_μ the dimensionality of the μth irrep. Then another powerful result states that

$$\sum_{\mu=1}^{\mu=c} n_\mu^2 = g.$$

5. The above relation restricts the possible dimensions of an irrep of G. For example, an abelian group ($c = g$) has *only one-dimensional irreps*. A nonabelian group ($c < g$) necessarily has at least one irrep of dimension > 1.

[46] 'irrep' is an accepted short-form for an irreducible representation.

6. An irrep is uniquely (up to equivalence) characterised by its characters. The characters of an irrep satisfy

$$\sum_{a \in G} |\chi(a)|^2 = g.$$

For a reducible representation, the left-hand side is necessarily greater than g.

7.4.6 Symmetry and quantum mechanics

In this subsection, we shall explain the important connection between quantum mechanics and group theory and the theory of group representations.

Let H be the Hamiltonian of a physical system. Let the system be invariant under a symmetry group G. For every symmetry operation $R \in G$, there is a unitary operator U_R acting on states of the system. Then the fact that R leaves the system invariant is translated as

$$U_R H U_R^{-1} = H, \quad \text{i.e.,} \quad [U_R, H] = 0.$$

Let ψ be a stationery state wave function: $H\psi = E\psi$. Then,

$$H(U_R\psi) = U_R H\psi = E(U_R\psi).$$

Thus the wave function $U_R\psi$ is also associated with the same energy E for any $R \in G$.

We now consider the set $\{U_R\psi;\ R \in G\}$ and count the number of *linearly independent* members of the set. Let one such set of linearly independent wave-functions be $\{\psi_1, \psi_2, \dots, \psi_d\}$. These transform under rotations as follows:

$$U_R\psi_i = \sum_{j=1}^{d} D(R)_{ij}\psi_j.$$

Then the set of matrices $\{D(R);\ R \in G\}$ forms a $d \times d$ representation of G. This implies a d-fold degeneracy. Barring accidential degeneracy, this representation will be irreducible. Thus, the energy levels of a physical system with a symmetry group G will be in (degenerate) groups of $n'_\mu s$, the dimensionalities of the irreducible representations of G.

An illustrative example

We shall now give an illustration of the results outlined earlier in this section. We shall consider a simple physical system, the ammonia molecule, NH_3. We shall study its symmetry properties and the symmetry group, study the irreps of the symmetry group, and draw conclusions about the energy level structure and other properties of stationary states just from the knowledge of the group theory. (These will eventually enable one to know about the allowed electronic transitions and their intensities – a topic beyond the scope of the book.) However, consider first the complexity of the physical system as such. It consists of 4 nuclei and 10 electrons interacting via a complex many-body interaction, including a spin–orbit interaction whose magnitude is a priori unknown. An exact solution to the many body problem would well neigh be impossible. However, as remarked in the introduction, while we cannot solve the problem by analytic methods, we can obtain the *structure* of energy levels and further knowledge about the stationary states from group theory considerations alone. This first requires us to determine the set of all symmetry operations of the physical system which we know forms a group.

We recall that the NH_3 molecule is a system with a tetrahedral structure. It has an N atom at the apex and three H-atoms placed symmetrically at the three vertices of an equilateral triangle which forms the base of the tetrahedron. The tetrahedron is not a regular one. We shall choose the coordinate system with origin at the center of the base and with the N atom on the z-axis. We shall also assume that the $x - z$ plane bisects the base. It is left as an exercise to verify that the following operations are symmetry operations of the system:

1. $C_3 =$ a rotation of $2\pi/3$ about the z-axis.
2. $\sigma \equiv$ a reflection in the $x - z$ plane

We note that the above operations satisfy $\sigma^2 = e = (C_3)^3$ and thus are elements of order 2 and 3 respectively. Of course, the linearly independent powers and products of these, viz., C_3^2, $C_3\sigma$ and $C_3^2\sigma$ must also be elements of the symmetry group. These, in fact, stand for a rotation through 240° about the z-axis and the reflection in planes bisecting the other two faces of the tetrahedron. Thus, these five operations together with the identity operation, e, form a nonabelian group, labelled earlier G_6, with the identifications $a = C_3$ and $b = \sigma$. We have also noted the class structure of the group:

$$\{e\}, \{a, a^2\}, \{b, ba, ba^2\},$$

and thus has $c = 3$ classes. Applying the theorems stated earlier, the group will have 3 irreps with dimensions n_μ given by $\sum_{\mu=1}^{3} n_\mu^2 = 6$. Thus, $n_1 = 1, n_2 = 1, n_3 = 2$. We have written down earlier specific matrices for the two-dimensional irrep of G_6 (see Section 7.4.1, Example 3). The other two one-dimensional irreps are easy to write down:

1. $D^{(1)}(c) = 1, \forall c \in G_6$ (the trivial irrep)
2. $D^{(2)}(c) = 1, c = e, a, a^2; D^{(2)}(c) = -1, c = b, ba, ba^2$

Now, we recall that since G_6 is a symmetry group of the system, every element of the group commutes with the Hamiltonian. Hence, the wave functions for the stationary states form multiplets of the group. In particular, we note the following results regarding the stationary states of the system:

1. The energy levels will belong to one of each of the three irreps.
2. Barring any accidental degeneracy of the system, an energy level belonging to one of the 1-dimensional irreps will be non-degenerate, while those belonging to the two-dimensional irrep will be two-fold degenerate.
3. The wave function of the state belonging to the trivial irrep $D^{(1)}$ will be invariant under all symmetry operations.
4. The wave function of the state belonging to the irrep $D^{(2)}$ will be invariant under the three rotations, while under any of the three reflections it will change sign (recall the values of $D^{(2)}(c)$ above).
5. The two wave functions of the states that belong to the two-dimensional irrep, will under the group operations involving the three reflections, mix with each other. This is in view of the fact that the representation matrices for $D(b)$, $D(ba)$, $D(ba^2)$ are non-diagonal.

A knowledge of the properties of the stationary state wave functions under symmetry operations, in fact, allows one to deduce much more: for example the selection rules for electronic transitions, etc. For this the reader is referred to the bibliography at the end of the book.

7.5 Continuous Groups

In physics, we come across many examples of groups where the group elements depend continuously on one or more parameters. For example, we know that the set of rotations of a rigid body about a fixed axis forms a group. Each element of the group can be parametrised by one parameter, (say) the rotation angle $\theta (0 \leq \theta < 2\pi)$, which varies continuously. Such a group is called a *one-parameter continuous group*. One can consider also the set of *all* rotations of a rigid body about a fixed point. This also forms a group (called $SO(3)$). Each element of the group can be labelled by a set of three continuously varying parameters $(\theta_1, \theta_2, \theta_3)$ and is thus a three-parameter group. Similarly, the set of all unitary operations consisting of the time evolution of a physical system forms a one-parameter group. The set of all space translations forms a three-parameter group. There are many other continuous groups that appear in physics. For example, in a relativistically invariant theory, there is the symmetry group of the set of all Lorentz transformations. This is a six-parameter group. Combining it with the four-parameter group of space–time translations, one comes to the 10-parameter Poincare group. Nuclear and particle physics use the $SU(2)$ group of rotations in the *isospin* space. Particle physicists find it useful to employ groups such as $SU(3)$ for the classification of elementary particles. The standard model of the fundamental interactions has the group $SU(3) \otimes SU(2) \otimes U(1)$ as the symmetry group of the action. Modern theories of elementary particle physics make use of several other continuous groups for the construction of models that attempt to go beyond the standard model.

7.5.1 An illustrative example

We shall illustrate the ideas underlying continuous groups by taking a simple example.

Consider the set of 2×2 real non-singular matrices

$$G = \{A_{2\times 2} : \det A \neq 0, a_{ij} \in \mathcal{R}\ \}$$

which forms a group under matrix multiplication called $GL(2, R)$ (see Exercise 7.18). A typical element A of G can be represented as

$$A = \begin{bmatrix} a & b \\ c & d \end{bmatrix},$$

where, a, b, c and d are real arbitrary numbers subject to $ad - bc \neq 0$. Clearly, A depends continuously on 4 real parameters (see however Exercise 7.20). G has an infinite number of elements and is thus an infinite group. Given $A, B \in G$, the commutator $[A, B]$ need not vanish (give an example!) and hence it is a nonabelian group. Consider now the identity element and a 'neighbouring' group elements (neighbouring in the sense that their difference has infinitesimal elements):

$$I = \begin{bmatrix} 1 & 0 \\ 0 & 1 \end{bmatrix}; \quad B = \begin{bmatrix} 1+\delta b_1 & \delta b_2 \\ \delta b_3 & 1+\delta b_4 \end{bmatrix} \equiv I + \delta B.$$

We can now express δB as

$$\delta B = \sum_{i=1}^{4} \delta b_i T_i$$

where T_i is a set of the four matrices given below.

$$T_1 = \begin{bmatrix} 1 & 0 \\ 0 & 0 \end{bmatrix}, \quad T_2 = \begin{bmatrix} 0 & 1 \\ 0 & 0 \end{bmatrix}$$
$$T_3 = \begin{bmatrix} 0 & 0 \\ 1 & 0 \end{bmatrix}, \quad T_4 = \begin{bmatrix} 0 & 0 \\ 0 & 1 \end{bmatrix} \tag{7.11}$$

It is easy to verify that the T's form a set of four real linearly independent 2×2 matrices. (In fact, the T's constitutes a basis for the expansion of any real 2×2 matrix). We note now that if we compute the commutator of any two T's, viz., $[T_i, T_j]$, then this itself being a real 2×2 matrix, it can be expanded as a linear combination of the T's as,

$$[T_i, T_j] = f_{ij}^k T_k. \tag{7.12}$$

The f_{ij}^k's above are called the 'structure constants' and are real numbers. A simple exercise allows one to obtain the structure constants $f_{ij}^k = -f_{ji}^k$ (see Exercise 7.16).

The T_i's are referred to as the *generators*[47] of the group and (7.12) is referred to as the Lie Algebra of the group.

We further note that *any* set of three $N \times N$ square matrices always identically satisfy

$$[[A, B], C] + [[B, C], A] + [[C, A], B] = 0, \tag{7.13}$$

which is easily verified by an explicit expansion and is called the 'Jacobi identity'. For the generators, we have,

$$[[T_i, T_j], T_l] + [[T_j, T_l], T_i] + [[T_l, T_i], T_j] = 0. \tag{7.14}$$

If we combine (7.12)with (7.14), we obtain

$$f_{ij}^k f_{kl}^m + f_{jl}^k f_{ki}^m + f_{li}^k f_{kj}^m = 0. \tag{7.15}$$

This is referred to as the 'Jacobi identity for the structure constants'.

We further note that the generators of the Lie group also serve as a basis for a real linear vector space of the set of all 2×2 matrices (not necessarily non-singular), which is to be distinguished from group G.

We also note that a particular combination of generators, viz.,

$$T_1 + T_4 = \begin{bmatrix} 1 & 0 \\ 0 & 1 \end{bmatrix},$$

being the identity matrix, commutes with all generators. We further note that the commutators of the remaining three independent linear combinations $T_1 - T_4, T_2, T_3$ (being traceless) do not involve $T_1 + T_4$.

The above prototype example demonstrates the essential elements associated with a continuous group. These are:

1. Presence of one or more continuous parameters
2. An infinite number of elements
3. A set of n generators $\{T_i, i = 1, 2, \ldots, n\}$ corresponding to the n independently varying continuous parameters. These generator T_i tell us how a group element varies with i^{th} parameter as in the vicinity of the identity element.
4. A Lie algebra of n generators:
$$[T_i, T_j] = f_{ij}^k T_k \tag{7.16}$$
5. Structure constants f_{ij}^k, antisymmetric in (i, j) and satisfying the Jacobi identity
6. n Generators that span a linear vector space (possibly over a complex field) of dimension n

[47] In mathematics literature, they are called the *infinitesimal* generators.

Solved problem 7.5 Consider the set of the following transformations

$$x' = a + x\cos\phi + y\sin\phi$$
$$y' = b + y\cos\phi - x\sin\phi \quad -\infty < a, b < \infty; \quad 0 \le \phi < 2\pi \tag{7.17}$$

(a) Show that the set of all such transformations form a group. Find the law of composition of parameters.
(b) Is the group abelian?
(c) Does the group have an invariant subgroup?

Solution:

(a) A typical transformation g is characterised by three parameters (a, b, ϕ). Consider applying two such transformations g_1 and g_2 with parameters (a_1, b_1, ϕ_1) and (a_2, b_2, ϕ_2) successively. g_1 transforms (x, y) to (x', y') given by

$$x' = a_1 + x\cos\phi_1 + y\sin\phi_1,$$
$$y' = b_1 + y\cos\phi_1 - x\sin\phi_1. \tag{7.18}$$

g_2 transforms (x', y') to (x'', y'') given by

$$x'' = a_2 + x'\cos\phi_2 + y'\sin\phi_2,$$
$$y'' = b_2 + y'\cos\phi_2 - x'\sin\phi_2. \tag{7.19}$$

We can now re-express (x'', y'') in terms of (x, y). We then find:

$$x'' = a_2 + a_1\cos\phi_2 + b_1\sin\phi_2 + x\cos(\phi_1 + \phi_2) + y\sin(\phi_1 + \phi_2)$$
$$y'' = b_2 + b_1\cos\phi_2 - a_1\sin\phi_2 + y\cos(\phi_1 + \phi_2) - x\sin(\phi_1 + \phi_2)$$

We thus see that a succession of two transformations of the set yields another transformation of the set, viz.,

$$x'' = A + x\cos\Phi + y\sin\Phi,$$
$$y'' = B + y\cos\Phi - x\sin\Phi, \tag{7.20}$$

with,

$$A \equiv a_2 + a_1\cos\phi_2 + b_1\sin\phi_2,$$
$$B \equiv b_2 + b_1\cos\phi_2 - a_1\sin\phi_2,$$
$$\Phi \equiv (\phi_1 + \phi_2). \tag{7.21}$$

Thus, $g_1, g_2 \in S \Rightarrow g_2 g_1 \in S$.
We further note that the transformation with $a = 0, \phi = 0$ is the identity transformation. To obtain the element g_2 as the inverse of g_1 we set $g_2 g_1 = e$. In other words, we set, in the above equations, $A = B = \Phi = 0$. These lead to,

$$\begin{aligned} a_2 &= -a_1 \cos\phi_2 - b_1 \sin\phi_2 \\ b_2 &= -b_1 \cos\phi_2 + a_1 \sin\phi_2 \\ \phi_2 &= -\phi_1 \end{aligned} \tag{7.22}$$

This determines the inverse element of a given element.

(b) If the group were abelian, the order in which two group elements are combined would not matter and thus, the parameters A, B and Φ of (7.21) would be symmetric under the exchange of suffixes $(1 \leftrightarrow 2)$. As this is not true, the group is nonabelian.

(c) The set of pure translations (i.e., $\phi = 0$), T, is an invariant subgroup. First, it is a subgroup because if $g_1, g_2 \in T$, so does $g_1 g_2^{-1}$ which is another translation. Also, it is easy to verify from (7.21) that a succession of a pure rotation g, a pure translation followed by an inverse rotation g^{-1} leads to a pure translation: $gTg^{-1} \in T$.

7.5.2 The orthogonal group

Let G be the set of all 3×3 real orthogonal matrices. As is easy to verify that, (see Exercise 7.20(d)), it forms a group under matrix multiplication and is called $O(3)$. Consider a typical matrix O belonging to $O(3)$. It has 9 real elements. These 9 elements are constrained by the orthogonality condition $O^T O = I$. Written out explicitly, these conditions are

$$O_{ij} O_{ik} = \delta_{jk} \qquad j, k = 1, 2, 3. \tag{7.23}$$

If we consider the variation of the nine parameters O_{ij}, they cannot be varied arbitrarily; they have to satisfy the conditions arising from (7.23), viz.,

$$\delta O_{ij} O_{ik} + O_{ij} \delta O_{ik} = 0. \tag{7.24}$$

On account of the symmetry of (7.24) in (j, k), these constitute six independent conditions. Thus, the *variations* δO_{ij} of these nine parameters are constrained by six linear homogeneous conditions (7.24). Thus, only three of these variations are therefore linearly independent.

Therefore, an element infinitesimally close to identity can be expressed as

$$O = I + \sum_{i=1}^{3} \delta\xi_i T_i \equiv i\delta\xi_i \tau_i, \tag{7.25}$$

where $\delta\xi_i$ are three independent parameter variations (we have not specified exactly how a typical element O is parametrised; it is not necessary at this stage). Substituting the O of (7.25) in $O^T O = I$, we obtain, to first order in the parameter variations $\delta\xi_i$,

$$\sum_{i=1}^{3} \delta\xi_i [T_i + T_i^T] = 0.$$

In view of the fact that $\delta\xi_i$ are independent, we learn that each of the T_i's must be a (real) antisymmetric matrix ($T_i^T = -T_i$).

Now the linear vector space of all 3×3 antisymmetric real matrices is 3-dimensional. As its basis, we can choose

$$T_1 = i\tau_1 = \begin{bmatrix} 0 & 0 & 0 \\ 0 & 0 & -1 \\ 0 & 1 & 0 \end{bmatrix}, \quad T_2 = i\tau_2 = \begin{bmatrix} 0 & 0 & 1 \\ 0 & 0 & 0 \\ -1 & 0 & 0 \end{bmatrix}, \tag{7.26}$$

$$T_3 = i\tau_3 = \begin{bmatrix} 0 & -1 & 0 \\ 1 & 0 & 0 \\ 0 & 0 & 0 \end{bmatrix} \tag{7.27}$$

We can alternately write this as

$$(T_i)_{jk} = i(\tau_i)_{jk} = \epsilon_{jik}. \tag{7.28}$$

We note that the T and τ satisfy

$$[T_i, T_j] = \epsilon_{ijk} T_k,$$
$$[\tau_i, \tau_j] = i\epsilon_{ijk}\tau_k. \tag{7.29}$$

The above commutators are easily verified either by direct evaluation of the commutators or by using (7.28) together with one of the ϵ-identities:

$$\epsilon_{\text{min}}\epsilon_{njp} - \epsilon_{pin}\epsilon_{njm} = \epsilon_{ijk}\epsilon_{mkp}$$

7.5.3 $SU(N)$ group

Another set of groups that occurs most frequently in physics is the $SU(N)$ group. The group $SU(2)$ occurs in connection with the rotations ($SO(3)$ and $SU(2)$ have the same Lie algebra) and is required to accommodate the spin-1/2 representation (and other half-odd integral spins) of the rotation group. The groups $SU(3)$ occurs in the classification of elementary particles and also in the fundamental theory of strong interactions. The groups $SU(4)$, $SU(5)$ have been employed in various attempts at grand unification of fundamental forces.

We shall introduce the group $SU(N)$ through its, what is called, the 'fundamental representation'. It consists of the set G of all $N \times N$ unitary, unimodular matrices which, as we have seen in Section 7.2, forms a group under matrix multiplication.

We shall first determine the number of independently varying parameters of the group. Consider a typical element U of G. It has N^2 complex elements and thus $2N^2$ real parameters. The elements of U are however subject to the conditions arising from the unitarity and the unimodular nature of U. They are,

$$U^\dagger U = I \Rightarrow U_{ij}^* U_{ik} = \delta_{jk} \quad j, k = 1, 2, \ldots N. \tag{7.30}$$

$$\det U = 1. \tag{7.31}$$

First consider (7.30). With $j = k$, we have N *real* equations. For $j > k$, we have $N(N-1)/2$ *complex* equations. We also note that the $N(N-1)/2$ complex equations that we get for $j < k$ are simply complex conjugates of the former ones and are hence not independent. Hence these constitute $N(N-1)$ real conditions. Thus, the condition of unitarity places $N(N-1) + N = N^2$ constraints on $2N^2$ real parameters[48]. Now, $U^\dagger U = I$ necessarily implies $|\det U| = 1$ but leaves the phase of $\det U$ arbitrary. Hence the second condition (7.31) which fixes the *phase* of $\det U$ constitutes *one* further condition. Thus a general element of the group G has $(N^2 - 1)$ independently varying parameters.

We shall now obtain the Lie algebra of the group. To this end, we shall first consider an element of G infinitesimally close to the identity: $U = I + \delta U$. The condition of unitarity implies,

$$(I + \delta U)^\dagger (I + \delta U) = I \Rightarrow \delta U^\dagger + \delta U = 0 \tag{7.32}$$

to the first order in the infinitesimals. The condition that U is unimodular implies (to the first order in infinitesimals),

$$\det (I + \delta U) \simeq 1 + \mathrm{tr}\,(\delta U) = 1 \Rightarrow \mathrm{tr}\,(\delta U) = 0. \tag{7.33}$$

Thus, $-i\delta U$ is a infinitesimal traceless Hermitian matrix. Now, consider a linear vector space of $N \times N$ traceless hermitian matrices. It is easy to verify that it is an $(N^2 - 1)$ dimensional (real) linear vector space. We can choose as its basis the following set of traceless hermitian matrices:

(a) A set of $N - 1$ diagonal matrices $H_1, H_2, \ldots, H_{N-1}$ of the form

$$H_1 = \frac{1}{2}\begin{bmatrix} 1 & 0 & 0 & & 0 \\ 0 & 0 & 0 & & 0 \\ 0 & 0 & 0 & & 0 \\ 0 & 0 & 0 & & 0 \\ 0 & 0 & 0 & & -1 \end{bmatrix} \quad H_2 = \frac{1}{2}\begin{bmatrix} 0 & 0 & 0 & & 0 \\ 0 & 1 & 0 & & 0 \\ 0 & 0 & 0 & & 0 \\ 0 & 0 & 0 & & 0 \\ 0 & 0 & 0 & & -1 \end{bmatrix}; \cdots \tag{7.34}$$

[48] How will you verify that these N^2 constraints are really independent?.

(b) and a set of $N(N-1)/2$ matrices given by

$$T^{ij}_{mn} = \frac{1}{2}[\delta_{im}\delta_{jn} + \delta_{in}\delta_{jm}] \quad i < j$$

(c) and a set of $N(N-1)/2$ matrices given by

$$T'^{ij}_{mn} = \frac{1}{2}[i\delta_{im}\delta_{jn} - i\delta_{in}\delta_{jm}] \quad i < j,$$

each of which is hermitian and traceless. (We can think of these as simple generalizations of $\frac{\sigma_3}{2}$, $\frac{\sigma_1}{2}$ and $\frac{\sigma_2}{2}$ respectively). We can then express the infinitesimal matrix δU as

$$\delta U = \sum_{1}^{N-1} c_i H_i + \sum_{i<j} a_{ij} T^{ij} + + \sum_{i<j} b_{ij} T'^{ij}. \tag{7.35}$$

The structure constants for the group $SU(N)$ can be found for the above expressions for the generators and the Lie algebra

$$[T_i, T_j] = i f^k_{ij} T_k.$$

This group has $(N-1)$ mutually commuting generators $H_1, H_2, \ldots, H_{N-1}$. The maximum number of mutually commuting generators is called the rank of a group. For $SU(N)$, it is $(N-1)$. We shall close the discussion by giving the algebra for the commonly occuring groups: $SU(2)$ and $SU(3)$. For $SU(2)$, we chose the generators as

$$H_1 = \frac{1}{2}\begin{bmatrix} 1 & 0 \\ 0 & -1 \end{bmatrix} \equiv \frac{1}{2}\sigma_3,$$

$$T_{12} = \frac{1}{2}\begin{bmatrix} 0 & 1 \\ 1 & 0 \end{bmatrix} \equiv \frac{1}{2}\sigma_1,$$

$$T'_{12} = \frac{1}{2}\begin{bmatrix} 0 & -i \\ i & 0 \end{bmatrix} \equiv \frac{1}{2}\sigma_2. \tag{7.36}$$

As we know, the generators $\tau_i = \frac{1}{2}\sigma_i$ satisfy the commutations rules

$$[\tau_i, \tau_j] = i\epsilon_{ijk}\tau_k.$$

This implies the following structure constants: $f^k_{ij} = \epsilon_{ijk}$. We note that these are totally antisymmetric. This is a general property of 'semi-simple' groups (as elaborated in the next subsection) that the generators can always be chosen such that the structure constants are totally antisymmetric.

Regarding $SU(3)$, it is conventional to choose the eight Gell-Mann matrices $\left\{\frac{1}{2}\lambda_i, i = 1, 2, \ldots, 8\right\}$ as the generators. They are:

$$\lambda_1 = \begin{bmatrix} 0 & 1 & 0 \\ 1 & 0 & 0 \\ 0 & 0 & 0 \end{bmatrix}; \quad \lambda_2 = \begin{bmatrix} 0 & -i & 0 \\ i & 0 & 0 \\ 0 & 0 & 0 \end{bmatrix}; \quad \lambda_3 = \begin{bmatrix} 1 & 0 & 0 \\ 0 & -1 & 0 \\ 0 & 0 & 0 \end{bmatrix};$$

$$\lambda_4 = \begin{bmatrix} 0 & 0 & 1 \\ 0 & 0 & 0 \\ 1 & 0 & 0 \end{bmatrix}; \quad \lambda_5 = \begin{bmatrix} 0 & 0 & -i \\ 0 & 0 & 0 \\ i & 0 & 0 \end{bmatrix};$$

$$\lambda_6 = \begin{bmatrix} 0 & 0 & 0 \\ 0 & 0 & 1 \\ 0 & 1 & 0 \end{bmatrix}; \quad \lambda_7 = \begin{bmatrix} 0 & 0 & 0 \\ 0 & 0 & -i \\ 0 & i & 0 \end{bmatrix}; \quad \lambda_8 = \frac{1}{\sqrt{3}}\begin{bmatrix} 1 & 0 & 0 \\ 0 & 1 & 0 \\ 0 & 0 & -2 \end{bmatrix}. \tag{7.37}$$

7.5.4 A summary of representations of *SO*(3), *SU*(2) and *SU*(3)

$SO(3)$: These representations occur in the theory of angular momentum and have an easily identifiable meaning. The finite-dimensional irreducible representations of $O(3)$ are all characterised by a single nonnegative integer l and are $(2l + 1)$-dimensional real representations. We identify the l-dimensional representation space with the vector space of the real traceless symmetric tensors of rank l, which is $(2l + 1)$-dimensional. Thus the representation matrices of the $l = 1$ representation, which are 3×3 matrices, give the effect of rotations on a 3-dimensional vector.

$SU(2)$: As mentioned earlier, the groups $SO(3)$ and $SU(2)$ have the same Lie algebra. In addition to the representations characterised by an integer $j = l$, now there are additional irreps characterised by a half odd integer j: $j = 1/2, 3/2, 5/2, \ldots$ and they are $(2j+1)$-dimensional complex representations.

$SU(3)$: These representations are found very useful in classifying the elementary particles. The irreducible representations are now characterised by two parameters (p, q) (both non-negative integers) and its dimensions, $D(p, q)$, is given by

$$D(p, q) = 1/2(p + 1)(q + 1)(p + q + 2).$$

The lowest non-trivial representation is of 3-dimensions called the fundamental representation.

Finally, we remark that there is always a representation of the dimension equal to the number of generators of a Lie group called the 'adjoint' representation. For this representation, the generators are represented by matrices constructed out of the structure constants: $T^i_{jk} = if_{jik}$.

Exercises

On § 7.2

7.1 Show that the set G of all 2×2 real matrices forms a group under matrix *addition*. What is now the identity element? Does the set of all 2×2 real *non-singular* matrices form a group under this 'law of multiplication'?

7.2 Let $a \in G$, a finite group. Show by considering the set

$$\{a, a^2, \ldots, a^{m-1}, a^m = e\}$$

that the order of any element $a \in G$, must be less than or equal to $O(G)$. Show further that it must be ≥ 2 except when $a = e$.

7.3 Consider the set $G = \{a, a^2, a^3, \ldots a^m \equiv e\}$ with $O(a) = m$. Show that this is a abelian group.

7.4 Let P be the parity (inversion) operator acting on the coordinates of a point: $P\mathbf{r} \equiv -\mathbf{r}$. Show that $G = (I, P)$ is a cyclic group of order 2.

7.5 Let R_x, R_y, R_z the three operations that correspond to three reflections in the (y, z),(z, x) and (x, y) planes respectively. Show that each of these elements are of order 2. Also, show that the smallest group containing all of these is

$$G = \{I, R_x, R_y, R_z, R_xR_y, R_yR_z, R_zR_x, P \equiv R_xR_yR_z\},$$

which is an abelian group of order 8. What does the element R_xR_y correspond to physically? Also, show that P corresponds to the operation of space inversion.

7.6 Let G be a finite group of order n. Let $a \in G$. Show directly that the order p of 'a' is a divisor of n. (This is alternately understood as a consequence of Lagrange's theorem). Do this in the following steps.

(i) Note that the elements of $C_p \equiv \{e, a, a^2, \ldots, a^{p-1}\}$ are all distinct.

(ii) If $p < n$, consider an element $b \in G$ such that such that $b \notin C_p$. Consider the set $bC_p \equiv \{b, ba, ba^2, \ldots, ba^{p-1}\}$. Show that all the elements of bC_p are all distinct and distinct from those of C_p.

(iii) If now $2p < n$, continue this process. Note that the process must end in finite number of steps.

7.7 Using the result of the previous problem, show that a group of prime order p must be cyclic.

7.8 Consider a rectangular parallelepiped of sides a, b and c, all distinct. Write down the set of all rotations about its centre that keeps the parallelepiped in its place (symmetry operations). Show that it is a group of order 4. Can you identify its multiplication table from the text? Now include all the reflections in the planes through the centre that are symmetry operations. Show now that the symmetry group is of order 8.

7.9 Write down the symmetry operations of a square that keep its centre fixed. Show that it is a group of order 8. Is it abelian?

On § 7.3

7.10 Verify that the condition (iii) employed in the alternate way of defining a subgroup (see Section 7.3) is in fact equivalent to conditions (i) and (ii) in its definition taken together.

7.11 Show that the set G of all $N \times N$ complex non-singular matrices forms a group under matrix multiplication. Show by giving an explicit example that this group is nonabelian. What is the order of the group? Show that the set of $N \times N$ real diagonal matrices forms an abelian subgroup of G. Can you suggest some other subgroups of G?

7.12 Let $G_1 = \{e_1, a_1, b_1, \ldots\ldots\}$ and $G_2 = \{e_2, a_2, b_2, \ldots\ldots\}$ be two groups and let $G = G_1 \otimes G_2$ be their direct product. Let

$$G_1' = \{(e_1, e_2), (a_1, e_2), (b_1, e_2), \ldots\ldots\ldots\}.$$

Show that G_1' is an invariant subgroup of G.

On § 7.4

7.13 Show that the set of all matrices taken from the rth blocks of the expressions of the form (7.9), viz., $D_{(r)}G \equiv \{A_r(a);\ a \in G\}$, constitutes a representation of G.

7.14 Show that

(a) different classes of G have no elements in common.
(b) the identity element e is a class by itself.
(c) for an abelian group, each class has only one element. For a finite abelian group, the number of classes 'c' is also the number of elements $g = O(G)$.
(d) for a finite nonabelian group, $c < g$.

7.15 Prove the following properties of the characters.

(a) Character is unchanged by a similarity transformation and hence the set of characters is the same for all equivalent representations.
(b) All elements in the same class have the same characters.
(c) For an $N \times N$ representation, $\chi(e) = \text{tr}I = N$.
(d) For a one-dimensional representation, $D(a) = \chi(a)$.
(e) For a $N \times N$ unitary representation, $|\chi(a)| \leq N$.

On § 7.5

7.16 Show that $T_i;\ i = 1, 2, 3, 4$ of (7.11), is a basis of a real linear vector space of 2×2 real matrices.

7.17 Obtain the structure constants f_{ij}^k's.

7.18 Show that if a new basis $T_i' = \sum_{j=1}^4 a_{ij} T_j$ with $\det A \neq 0$ is chosen for the generators, then f_{ij}^k's are replaced by

$$f_{ij}^{\prime k} = \sum_1^4 \sum_1^4 a_{im} a_{jm} f_{mn}^p (A^{-1})_p^k.$$

7.19 Verify the identity of (7.29)): $[T_i, T_j] = \epsilon_{ijk} T_k$.

7.20 Show that the set of all 2 × 2 real non-singular matrices forms a group under matrix multiplication. This group is called $GL(2, R)$. Show further that the set of all 2 × 2 real non-singular matrices with determinant one is an invariant subgroup of it called $SL(2, R)$. Show that the latter is a 3-parameter group. Write down the generators of $SL(2, R)$ and its Lie algebra. Show that both of these groups are nonabelian.

7.21 Consider the matrix $A = \begin{bmatrix} a & b \\ c & d \end{bmatrix}$ with $ad - bc \neq 0$. Consider $b = c = d = 1 =$ fixed and let a be a variable with $a \neq 1$ (so that $ad - bc \neq 0$).

(a) Show that given $b = c = d = 1$, a cannot be varied continuously across the value $a = 1$.

(b) Show however that for any (a_0, b_0, c_0, d_0) such that $a_0 d_0 - b_0 c_0 \neq 0$, there is some nonzero interval for each parameter which includes (a_0, b_0, c_0, d_0) in which each of a, b, c, d can be varied continuously.

(c) Show therefore that the observation in (a) above does not invalidate the discussion in Section 7.5.1.

(d) Show that the set of 3 × 3 real orthogonal matrices forms a group (called $O(3)$), and the subset of this set with unit determinant forms a subgroup (called $SO(3)$).

References

[1] Hamermesh, M., *Group Theory and its Application to Physical Problems*, Addison-Wesley Pub. Co, Reading, Mass., 1962.

[2] Tinkham, Michael, *Group Theory and Quantum Mechanics*, McGraw-Hill, New York, 1964.

[3] Wigner, Eugene Paul, *Group Theory and its Application to the Quantum Mechanics of Atomic Spectra*, Academic Press, New York, 1959.

[4] Gilmore, Robert, *Lie Groups, Lie Algebras, and some of their Applications*, Krieger Pub. Co, Malabar, Fla, 1994.

[5] Wybourne, Brian, G., *Classical Groups for Physicists*, Wiley, New York, 1974.

Appendix

In this appendix we shall summarise the rigorous results that are required in order to interchange the order of various mathematical operations. For proofs, the reader is directed to standard books on analysis[49].

1. **Differentiation of an infinite summation**
   Let $\sum_{n=0}^{\infty} f_n(z)$ be a series such that (i) it converges uniformly along a contour C; (ii) $f_n(z)$ is analytic throughout C and its interior. Then (i) the sum converges throughout the interior of C, (ii) it is an analytic function in the interior of C and (iii) it can be differentiated term by term.
   In particular, consider a power series $\sum_0^{\infty} c_n z^n$ within its circle of convergence $|z| = R$. There, it can be differentiated term by term.
2. **Differentiation under a definite integral**
   The equation

$$\frac{d}{dk}\int_a^b dx f(x,k) = \int_a^b dx \frac{d}{dk} f(x,k) \qquad (-\infty < a < b < \infty),$$

   is true if

   (a) $\int_a^b dx f(x,k)$ exists, and

   (b) $\frac{df}{dk}$ is a continuous function of both the variables (x, k).

3. **Differentiation under an infinite integral**
   The equation,

$$\frac{d}{dk}\int_a^{\infty} dx f(x,k) = \int_a^{\infty} dx \frac{d}{dk} f(x,k)$$

   is true at $k = k_0$ where k_0 is contained in some neighbourhood (c, d) if

   (a) $\int_a^{\infty} dx f(x,k)$ exists, and

---

[49] See for example, *A Course on Modern Analysis* by Whittaker and Watson, and *Mathematical Analysis* by Apostel.

(b) the infinite integral $\int_a^\infty dx \frac{d}{dk} f(x, k)$ converges uniformly whenever k lies in the interval (c, d),

(c) $f(x, k)$ is a continuous function of both the variables (x, k) for $x \geq a$ and $k \in (c, d)$.

Similar result applies to $\int_{-\infty}^{a} dx f(x, k)$. The two can be put together for an integral $\int_{-\infty}^{\infty} dx f(x, k)$.

4. **Limit under an integration**

Let $I(k) = \int_a^\infty dx f(x, k)$ converge uniformly for all $k \in (c, d)$. Then, if $f(x, k)$ is a continuous function of both variables (x, k) for $x \geq a$ and $k \in (c, d)$, then $I(k)$ is continuous in k when $k \in (c, d)$.

5. **Interchange of order of integration**

Let $f(x, k)$ be a piecewise continuous function in each of the variables (x, k). Now, consider the two integrals

$$I_1 = \int_{-\infty}^{\infty} dx \int_{-\infty}^{\infty} dk f(x, k)$$

and

$$I_2 = \int_{-\infty}^{\infty} dk \int_{-\infty}^{\infty} dx f(x, k).$$

If either of the two integrals exist, then $I_1 = I_2$.

Bibliography

[1] Courant, Richard and Hilbert, David, *Methods of Mathematical Physics*, Interscience Publishers, New York, 1989.

[2] Morse, Philip McCord and Feshbach, Herman, *Methods of Theoretical Physics, Vol. I–II*, McGraw-Hill, Boston, Mass., 1999.

[3] Whittaker, E. T. and Watson, G. N., *A Course of Modern Analysis: An Introduction to the General Theory of Infinite Processes and of Analytic Functions, with an Account of the Principal Transcendental Functions*, AMS Press, New York, 1979.

[4] Denenry, P. and Krzywicki, A., *Mathematics for Physicists*, Harper and Row, New York, 1967.

[5] Arfken, George, B. and Weber, Hans-Jurgen, *Mathematical Methods for Physicists*, Harcourt/Academic Press, San Diego, 2001.

[6] Brown, James Ward and Churchill, Ruel Vance, *Complex Variables and Applications*, McGraw-Hill Higher Education, Boston, 2004.

Index